SpringerBriefs in Optimization

SpringerBriefs present concise summaries of cutting-edge research and practical applications across a wide spectrum of fields. Featuring compact volumes of 50 to 125 pages, the series covers a range of content from professional to academic. Briefs are characterized by fast, global electronic dissemination, standard publishing contracts, standardized manuscript preparation and formatting guidelines, and expedited production schedules.

Typical topics might include:

- A timely report of state-of-the art techniques
- A bridge between new research results, as published in journal articles, and a contextual literature review
- A snapshot of a hot or emerging topic
- An in-depth case study
- A presentation of core concepts that students must understand in order to make independent contributions

SpringerBriefs in Optimization showcase algorithmic and theoretical techniques, case studies, and applications within the broad-based field of optimization. Manuscripts related to the ever-growing applications of optimization in applied mathematics, engineering, medicine, economics, and other applied sciences are encouraged.

Titles from this series are indexed by Web of Science, Mathematical Reviews, and zbMATH.

More information about this series at http://www.springer.com/series/8918

V. A. Kalyagin • A. P. Koldanov
P. A. Koldanov • P. M. Pardalos

Statistical Analysis of Graph Structures in Random Variable Networks

 Springer

V. A. Kalyagin
Laboratory of Algorithms and Technologies
for Networks Analysis
National Research University
Higher School of Economics
Nizhny Novgorod, Russia

P. A. Koldanov
Laboratory of Algorithms and Technologies
for Networks Analysis
National Research University
Higher School of Economics
Nizhny Novgorod, Russia

A. P. Koldanov
Laboratory of Algorithms and Technologies
for Networks Analysis
National Research University
Higher School of Economics
Nizhny Novgorod, Russia

P. M. Pardalos (iD)
Department of Industrial & Systems
Engineering
University of Florida
Gainesville, FL, USA

ISSN 2190-8354   ISSN 2191-575X (electronic)
SpringerBriefs in Optimization
ISBN 978-3-030-60292-5 ISBN 978-3-030-60293-2 (eBook)
https://doi.org/10.1007/978-3-030-60293-2

This Springer imprint is published by the registered company Springer Nature Switzerland AG
The registered company address is: Gewerbestrasse 11, 6330 Cham, Switzerland

Preface

Network analysis in our days is a rapidly developing area of analysis of complex systems. A complex system is understood as a network of a large number of interacting elements. Examples of such systems are the stock market, the gene expression network, brain networks, the climate network, the Internet, social networks, and many others. Network analysis of such complex systems involves the construction of a weighted graph, where the nodes are the elements of the system, and the weights of the edges reflect the degree of similarity between the elements. To identify key information in a complex system, one can filter the huge amount of data to obtain representative unweighted subgraphs of the network. Such unweighted subgraphs, called network structures, include, in particular, the maximum spanning tree, the threshold graph, its cliques and independent sets, and other characteristics known in graph and network theory.

In this book we study complex systems with elements represented by random variables. Data analysis related with these systems is confronted with statistical errors and uncertainties. The main goal of this book is to study and compare uncertainty of algorithms of network structure identification with applications to market network analysis. For this we introduce a mathematical model of random variable network, define uncertainty of identification procedure through a risk function, discuss random variables networks with different measures of similarity (dependence), and study general statistical properties of identification algorithms. In addition, we introduce a new class of identification algorithms based on a new measure of similarity and prove its robustness in a large class of distributions. This monograph can be used by experts in the field and it can be helpful for graduate and PhD students. The work is supported by the lab LATNA in the framework of Basic Research Program at the National Research University Higher School of Economics and by RRFI grant 18-07-00524.

Nizhny Novgorod, Russia   Valery A. Kalyagin
Nizhny Novgorod, Russia   Alexander P. Koldanov
Nizhny Novgorod, Russia   Petr A. Koldanov
Gainesville, FL, USA   Panos M. Pardalos

Contents

Chapter 1
Introduction

Abstract Network analysis is widely used in modern data mining techniques. In this book, we consider networks where the nodes of a network are associated with random variables and edges reflect some kind of similarity among them. Statistical analysis of this type of networks involves uncertainty. This aspect is not well covered in existing literature. The main goal of the book is to develop a general approach to handle uncertainty of network structure identification. This approach allows to study general statistical properties (unbiasedness, optimality) of different identification algorithms and to design a new class of robust identification algorithms. Large area of applications varies from market network analysis to gene network analysis.

1.1 Network Structures

Consider a complex system with N elements. Network model of a complex system is a complete weighted graph with N nodes where the weights of edges reflect the degree of interaction between elements of the system. To identify key information in a complex system, one can reduce the complete weighted graph to simpler graph models. We will call such graph models *network structures*. There is a variety of network structures considered in the literature: concentration graph, threshold (market) graph, cliques and independent sets of the threshold graph, maximum spanning tree, planar maximally filtered graph, and others. *Concentration graph* is an unweighted graph obtained from the complete weighted graph by a simple filtration procedure. An edge is included in the concentration graph if and only if its weight in the complete weighted graph is nonzero. In a similar way, an edge is included in the *threshold graph* if and only if its weight is larger than a given threshold. *Clique* in a graph is a set of nodes such that every two nodes in the set are connected by an edge. *Independent set* is a set of nodes in a graph without edges between them. *Maximum spanning tree* in a weighted graph is the spanning tree of maximal total weight. *Planar maximally filtered graph* is the planar subgraph of a complete weighted graph of maximal total weight.

© The Author(s) 2020

V. A. Kalyagin et al., *Statistical Analysis of Graph Structures in Random Variable Networks*, SpringerBriefs in Optimization,
https://doi.org/10.1007/978-3-030-60293-2_1

The concentration graph gives information about the topology of pairwise connections in a complex system. The family of threshold graphs gives information about the variation of topology of pairwise connections with respect to the variable threshold. Cliques in the threshold graph are sets of closely connected elements of a complex system. Independent sets in the threshold graph are sets of nonconnected elements of a complex system. Maximum spanning tree and planar maximally filtered graph allow to detect a hierarchical clusters structure in a complex system.

1.2 Market Networks

Network approach has become very popular in the stock market analysis. Different network structures are considered. Maximum spanning tree (MST) was used in [66] to describe the hierarchical structure in USA stock market. This approach was further developed in [73] for portfolio analysis and in [74] for clustering in the stock market. Planar maximally filtered graph (PMFG) was introduced in [87], and it was used in [83] to develop a new clustering technique for the stock market. MST for different stock markets was investigated in [7, 8, 24, 26, 48, 69, 92].

Investigation of stock markets is using another popular network structure, the threshold or market graph, which was started in [5] and developed in [4, 6]. It was shown that cliques and independent sets of a threshold graph contain useful information on the stock market. Different stock markets using threshold (market) graph technique are investigated in [14, 27, 30, 39, 42, 71, 89].

Other network characteristics of the stock market attracted important attention in the literature too. The most influential stocks related with the stock index are investigated in [22, 23, 33, 81, 85]. Connections with random matrix theory were studied in [46, 70, 72, 76, 94]. Clustering and dynamics of a market network was investigated in a large number of publications, see for example [28, 40, 52, 53, 80, 84]. Big data aspects for market networks are developed in [31, 50, 51]. Network model for interbank connections was considered in [63]. Different measures of similarity for stock market networks were considered in [47, 82, 93, 95]. Review on the subject and large bibliography is presented in [67].

Most of publications are related with numerical algorithms and economic interpretations of obtained results. Much less attention is paid to uncertainty of obtained results generated by the stochastic nature of the market. The new approach to this problem is proposed in [44].

1.3 Gene Networks

According to Anderson [1], the use of graphical diagrams in genetics goes back to the work of the geneticist Sewall Wright (1921, 1934). In our days, " 'a graphical model in statistics is a visual diagram in which observable variables are identified with points (vertices or nodes) connected by edges and an associated family of probability distributions satisfying some independence specified by the visual pattern' " (see [1]). Fundamental probabilistic theory of Gaussian graphical models was developed in [59] where conditional independence is used for visualization. Some applications of graphical models for genetic analysis are presented in [58]. One of the important structures in graphical models is the concentration graph for the network where the weights of edges are given by partial correlations. Such a network is usually called gene expression network. In practice for gene expression network it is important to identify the concentration graph from observations. This problem is called *graphical model selection problem*. Different graphical model selection (identification of concentration graph) algorithms were developed in the literature. Detailed review on the topics is presented in [17].

Another type of gene network is so-called gene co-expression network. The concept of gene co-expression network was introduced in [12] and it is actively used in bioscience. An important structure for the gene co-expression network is the threshold graph for different measures of similarity (Pearson's correlation, Mutual Information, Spearman's rank correlation, and others). Weighted gene co-expression network analysis (WGCNA) was introduced in [96]. WGCNA is a tool for constructing and analyzing weighted gene co-expression networks with the scale-free topology. Detailed review on the subject can be found in [38].

Most of publications on gene networks are related with biological applications and interpretation of obtained results. Much less attention is paid to uncertainty of obtained results generated by stochastic nature of the network.

1.4 Uncertainty

To handle uncertainty, we introduce a concept of Random Variable Network (RVN), define a true network structure, and consider identification algorithms as statistical procedures. Then we define uncertainty by a risk function. This allows us to compare uncertainty of different identification procedures and study its general statistical properties, such as unbiasedness, optimality, and robustness (distribution free risk functions).

Random variable network is a pair (X, γ), where $X = (X_1, X_2, \ldots, X_N)$ is a random vector and γ is a measure of similarity (association) between a pair of random variables. For Gaussian Pearson correlation network, the vector X has multivariate Gaussian distribution and the measure of similarity γ is the Pearson correlation. In the same way, one can define the Student Pearson correlation network, the Gaussian partial correlation network, and so on. The random variable network generates a *network model*, a complete weighted graph with N nodes, where the weight of edge (i, j) is given by $\gamma_{i,j} = \gamma(X_i, X_j)$, $i \neq j$, $i, j = 1, 2, \ldots, N$. Note that we consider the symmetric pairwise measures only such that $\gamma_{i,j} = \gamma_{j,i}$. We will call this complete weighted graph *true network model*. The network structure in the true network model will be called *true network structure*.

We show in the book that different random variable networks can generate the same true network model and the same true network structures. This is the case, for example, where two different random vectors have elliptically contoured distributions and the measure of similarity γ is the Pearson correlation. Moreover, different random variable networks can generate functionally related true network models. This is the case, for example, where X has a given multivariate Gaussian distribution and two measures of similarity are Pearson and tau-Kendall correlations. This result is important for the adequate choice of measures of similarity for practical problems.

In practice, it is important to identify a true network structure from observations. We call this problem *network structure identification problem*. Any identification algorithm in this setting can be considered as statistical procedure which is a map from the sample space to the set of graphs with specific properties. For example, the threshold graph identification procedure is a map from the sample space to the set of simple unweighted graphs with N nodes, and MST identification procedure is a map from the sample space to the set of spanning trees with N nodes. We call *sample structure* the network structure obtained by an identification procedure.

Network structure identification uncertainty is defined by a difference between true and sample network structure. This difference generates a loss from a false decision. Expected value of the loss function is called risk function [62, 91]. We will use the risk function to measure uncertainty of network structure identification procedures. In general, the loss function can be associated with different distances between two graphs (true and sample network structures) such as symmetric difference, mutual information, K-L divergence, and others.

The network structure identification problem can be considered as multiple testing problem of individual hypotheses about edges. Two types of errors of identification arise: Type I error is the false inclusion of an edge in the structure (edge is absent in the true structure but it is present in the sample structure), Type II error is false exclusion of an edge from the structure (edge is present in the true structure, but it is absent in the sample structure). Well-known quality characteristics of multiple hypotheses testing procedures such as FWER (Family Wise Error Rate = probability of at least one Type I error) and FDR (False Discovery Rate) are particular case of the risk function for appropriate choice of the losses.

In terms of machine learning, the network structure identification problem is a binary classification problem. Each edge in the network model is classified in one of two classes: class Yes, if the edge is included in the true network structure, and class No if the edge is not included in the true network structure. In this setting, Type I error is associated with False Positive decision, and Type II error is associated with False Negative decision. Well-known characteristics of binary classification such as TPR (True Positive Rate) and FPR (False Positive Rate) are a particular case of the risk function for appropriate choice of the losses.

In our opinion, it is natural in network structure identification to consider the so called additive losses, i.e., the loss of the false decision about network structure is a sum of losses of the false decisions about individual edges. We show in the book that under some additional conditions, the risk function for additive losses is a linear combination of expected value of the numbers of Type I (FP) and Type II

(FN) errors. For this risk function, we study general statistical properties of known identification procedures. In particular, we show that the statistical procedure for the concentration graph identification based on individual partial correlation tests in Gaussian partial correlation network is optimal in the class of unbiased multiple testing procedures.

In practice the distribution of random vector X is unknown. Therefore, it is important to construct robust (distribution free risk function) network structure identification procedures. Such a procedure has a risk function which does not depend on distribution from a large class. We introduce in the book a new class of identification algorithms based on a new measure of similarity and prove its robustness (distribution free risk function) in the class of elliptically contoured distributions. This result can be applied in a wide area of practical applications.

1.5   Related Type of Networks

There are different types of networks, related with this book. In particular, functional brain networks and climate networks are closely connected with networks, considered in the book. In brain network one considers symmetrical measures of statistical association or functional connectivity – such as correlations, coherence, and mutual information – to construct undirected graphs and study their properties. A survey on this topics is presented in [11], see also [56]. In climate network, one uses a correlation between climate observations at different points and studies a threshold graph generated by this network. More detailed information can be found in [86]. Other types of networks are presented in the recent surveys [10, 97]. In the present book, we develop a general approach to study statistical properties of network structure identification algorithms and to measure their uncertainty. Most applications are given to market networks, but this approach can be developed for other networks too.

Chapter 2
Random Variable Networks

Abstract In this chapter, we give basic definitions related to random variable networks. After rigorous definitions of a random variable network and a network model, we give definitions of specific network structures: maximum spanning tree, planar maximally filtered graph, concentration graph, threshold graph, maximum clique, and maximum independent set in the threshold graph. All definitions are illustrated by examples. Then we consider a large class of network models generated by different distributions and different measures of similarity. For distributions, we use an important class of elliptical distributions. For pairwise measures of similarity, we consider Pearson correlation, Kruskal correlation, sign similarity, Fechner correlation, Kendall and Spearman correlations. As a main result of the chapter, we establish connections between network structures in different random variable network models. In addition, we define the partial correlation network and establish its properties for elliptical distributions.

2.1 Basic Definitions and Notations

Random variables network is a pair (X, γ), where $X = (X_1, \ldots, X_N)$ is a random vector, and γ is a pairwise measure of similarity (dependence, association,...) between random variables. One can consider different random variable networks associated with different distributions of the random vector X and different measures of similarity γ. For example, the Gaussian Pearson correlation network is the random variable network, where X has a multivariate Gaussian distribution and γ is the Pearson correlation. On the same way one can consider the Gaussian partial correlation network, the Gaussian Kendall correlation network, the Student Pearson correlation network, and so on.

The random variables network generates a network model. Network model for random variable network (X, γ) is the complete weighted graph with N nodes (V, Γ), where $V = \{1, 2, \ldots, N\}$ is the set of nodes, $\Gamma = (\gamma_{i,j})$ is the matrix of weights, $\gamma_{i,j} = \gamma(X_i, X_j)$. *Network structure* in the network model (V, Γ) is an unweighted graph (U, E), where $U \subset V$, E is the set of edges between nodes in

V. A. Kalyagin et al., *Statistical Analysis of Graph Structures in Random Variable Networks*, SpringerBriefs in Optimization,
https://doi.org/10.1007/978-3-030-60293-2_2

U. In this book, we consider the following network structures: maximum spanning tree (MST), planar maximally filtered graph (PMFG), concentration graph (CG), threshold graph (TG), maximum clique (MC), and maximum independent set (MIS) in the threshold graph.

The spanning tree in the network model (V, Γ) is a connected graph (network structure) (V, E) without cycles. Weight of the spanning tree (V, E) is the sum of weights of its edges $\sum_{(i,j) \in E} \gamma_{i,j}$. *Maximum spanning tree* (MST) is the spanning tree with maximal weight. Natural extension of maximum spanning tree is the *planar maximally filtered graph* (PMFG), planar connected graph (V, E) of maximal weight (see [29] for planar graph definition). There are different algorithms to find MST or PMFG in a weighted graph. In this book, we use well known Kruskal algorithm for MST and Kruskal type algorithm for PMFG. Both algorithms have a polynomial computational complexity.

The *concentration graph* in the graph model (V, Γ) is the network structure (V, E), where edge $(i, j) \in E$ if and only if $\gamma_{i,j} \neq 0$. *Threshold graph* in the network model (V, Γ) is the network structure (V, E), where $(i, j) \in E$ if and only if $\gamma_{i,j} > \gamma_0$ and γ_0 is a given threshold. Depending on the value of the threshold γ_0, threshold graph is varying from complete unweighted graph to the graph with isolated vertices. *Clique* in a threshold graph $G = (V, E)$ is the set of nodes $U, U \subset V$ such that $(i, j) \in E, \forall i, j \in U, i \neq j$. *Independent set* in a threshold graph $G = (V, E)$ is the set of nodes $U, U \subset V$ such that $(i, j) \notin E, \forall i, j \in U, i \neq j$. Maximum clique and maximum independent set problems are known to be NP-hard. In our computations, we use fast exact algorithm by Carraghan and Pardalos [15].

All these defined structures are popular in applications. MST is widely used in stock market network analysis. PMFG was used in market network analysis for cluster structure detection. Concentration graph gives information about a dependence structure in the network model (V, Γ). Family of threshold graphs gives information about the variation of topology of pairwise connections with respect to the variable threshold. It was observed [6] that for some threshold values the market graph has a scale free property, i.e., vertex degree distribution in the market graph follows a power law. Cliques in the threshold graph are sets of closely connected elements of the network model (V, Γ). For some markets (e.g., Russian market), maximum cliques are shown to be the most influential stocks of the market [89]. Independent sets in the threshold graph are sets of nonconnected elements of the network model. Maximum independent sets are known to be useful for portfolio optimization [43]. Maximum spanning tree and planar maximally filtered graph allow to detect a hierarchical clusters structure in the network model [88].

Example 2.1 Let us illustrate the introduced network structures by the following example. Consider network model with 10 nodes $V = \{1, 2, \ldots, 10\}$, and matrix Γ given by matrix below. The network structures are given by Figs. 2.1, 2.2, 2.3, 2.4 and 2.5. The Fig. 2.1 represents the maximum spanning tree. One can observe two clusters in MST (1, 2, 3, 4, 10 and 5, 6, 7, 8, 9) with the centers 9 and 10 connected by an edge. The Fig. 2.2 represents the planar maximally filtered

Fig. 2.1 Maximum spanning
tree for the network model of
the example 2.1

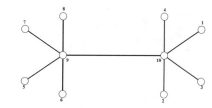

Fig. 2.2 Planar maximally
filtered graph for the network
model of the example 2.1

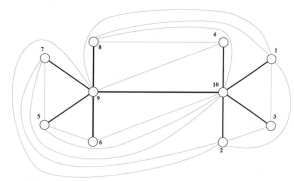

Fig. 2.3 Threshold graph for
the network model of the
example 2.1. Threshold
$\gamma_0 = 0.3$

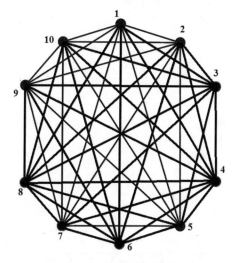

graph. The structure of PMFG is more complicated than MST. The filtered graph
contains more information about the structure of the considered network model,
new connections appear. There are 5 new connections inside clusters and 10 new
connections between clusters.

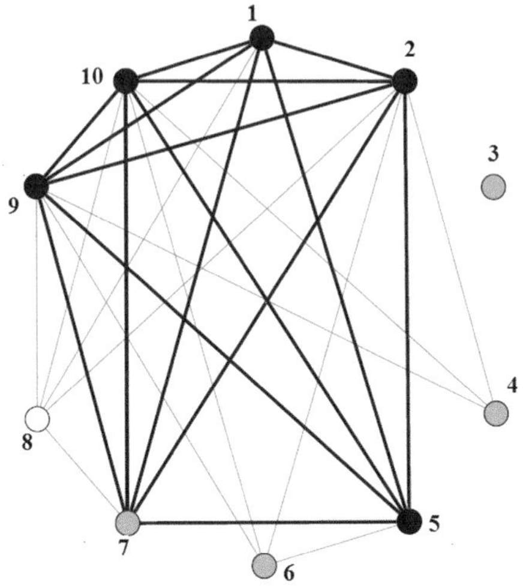

Fig. 2.4 Threshold graph for
the network model of the
example 2.1. Threshold
$\gamma_0 = 0.55$

Fig. 2.5 Threshold graph for
the network model of the
example 2.1. Threshold
$\gamma_0 = 0.7$

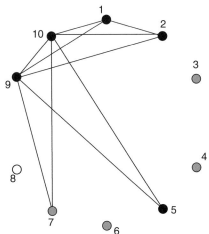

$$\begin{pmatrix} & 1 & 2 & 3 & 4 & 5 & 6 & 7 & 8 & 9 & 10 \\ 1 & 1.0000 & 0.7220 & 0.4681 & 0.4809 & 0.6209 & 0.5380 & 0.6252 & 0.6285 & 0.7786 & 0.7909 \\ 2 & 0.7220 & 1.0000 & 0.4395 & 0.5979 & 0.6381 & 0.5725 & 0.6666 & 0.6266 & 0.8583 & 0.8640 \\ 3 & 0.4681 & 0.4395 & 1.0000 & 0.3432 & 0.3468 & 0.2740 & 0.4090 & 0.4016 & 0.4615 & 0.4832 \\ 4 & 0.4809 & 0.5979 & 0.3432 & 1.0000 & 0.4518 & 0.4460 & 0.4635 & 0.4940 & 0.6447 & 0.6601 \\ 5 & 0.6209 & 0.6381 & 0.3468 & 0.4518 & 1.0000 & 0.5640 & 0.5994 & 0.5369 & 0.7170 & 0.7136 \\ 6 & 0.5380 & 0.5725 & 0.2740 & 0.4460 & 0.5640 & 1.0000 & 0.4969 & 0.4775 & 0.6439 & 0.6242 \\ 7 & 0.6252 & 0.6666 & 0.4090 & 0.4635 & 0.5994 & 0.4969 & 1.0000 & 0.6098 & 0.7161 & 0.7158 \\ 8 & 0.6285 & 0.6266 & 0.4016 & 0.4940 & 0.5369 & 0.4775 & 0.6098 & 1.0000 & 0.6805 & 0.6748 \\ 9 & 0.7786 & 0.8583 & 0.4615 & 0.6447 & 0.7170 & 0.6439 & 0.7161 & 0.6805 & 1.0000 & 0.9523 \\ 10 & 0.7909 & 0.8640 & 0.4832 & 0.6601 & 0.7136 & 0.6242 & 0.7158 & 0.6748 & 0.9523 & 1.0000 \end{pmatrix}$$

The Figs. 2.3, 2.4, and 2.5 represent the threshold graphs constructed for the thresholds $\gamma_0 = 0.3; 0.55; 0.7$. For $\gamma_0 = 0.3$, this graph is almost complete, only edge $(3, 6)$ is absent. There are two maximum cliques: $\{1,2,3,4,5,7,8,9,10\}$ and $\{1,2,4,5,6,7,8,9,10\}$. The maximum independent set is $\{3,6\}$. For $\gamma_0 = 0.55$, there are two maximum cliques with 6 vertices. One of them $\{1,2,5,7,9,10\}$ has the maximal weight. There are four maximum independent sets with 4 vertices. One of them $\{3, 4, 6, 7\}$ has minimal weight. For $\gamma_0 = 0.7$, the threshold graph has only 10 edges. There is one maximum clique with 4 vertices $\{1,2,9,10\}$. There are two maximum independent sets with 7 vertices $\{2,3,4,5,6,7,8\}$, $\{1,3,4,5,6,7,8\}$. Note that the described structures (MST, PMFG, MG, MC, MIS) are unweighted subgraphs of the network model and reflect different aspects of pairwise connections.

Remark This network model is taken from USA stock market. The following stocks are considered: A (Agilent Technologies Inc), AA (Alcoa Inc), AAP (Advance Auto Parts Inc), AAPL (Apple Inc), AAWW (Atlas Air Worldwide Holdings Inc), ABAX (Abaxis Inc), ABD (ACCO Brands Corp), ABG (Asbury Automotive Group Inc), ACWI (iShares MSCI ACWI Index Fund), ADX (Adams Express Company). Here $\gamma_{i,j}$ are the sample Pearson correlations between daily stock returns calculated by the 250 observations started from November 2010. Note that for another number of observations, one can obtain another value of sample Pearson correlations and consequently another network structures. This leads to uncertainty in network structure identification. One can construct a huge number of different network models of such type depending on the period of observations and measure of similarity used. The main questions is: how reliable are the results of network structures construction? Is there a stable structure behind these observations? How large is a deviation from this structure due to the stochastic nature of observations? To answer these questions, we propose a new approach based on the notion of random variable network.

2.2   Distributions and Measures of Similarity

In this section, we consider more in details different random variable networks (X, γ) associated with different distributions of the random vector X and different measures of similarity γ. We consider a large class of distributions, elliptically contoured distributions (or simply elliptical distributions), which are known to be useful in many applications [32]. Random vector X belongs to the class of elliptically contoured distributions if its density function has the form [1]:

$$f(x; \mu, \Lambda) = |\Lambda|^{-\frac{1}{2}} g\{(x - \mu)' \Lambda^{-1} (x - \mu)\} \tag{2.1}$$

where $\Lambda = (\lambda_{i,j})_{i,j=1,2,\dots,N}$ is positive definite symmetric matrix, $g(x) \geq 0$, and

$$\int_{-\infty}^{\infty} \cdots \int_{-\infty}^{\infty} g(y'y) dy_1 dy_2 \cdots dy_N = 1$$

This class includes in particular multivariate Gaussian distribution

$$f_{Gauss}(x) = \frac{1}{(2\pi)^{p/2} |\Lambda|^{\frac{1}{2}}} e^{-\frac{1}{2}(x-\mu)' \Lambda^{-1}(x-\mu)}$$

and multivariate Student distribution with ν degree of freedom

$$f_{Student}(x) = \frac{\Gamma\left(\frac{\nu+N}{2}\right)}{\Gamma\left(\frac{\nu}{2}\right) \nu^{N/2} \pi^{N/2}} |\Lambda|^{-\frac{1}{2}} \left[1 + \frac{(x-\mu)' \Lambda^{-1}(x-\mu)}{\nu} \right]^{-\frac{\nu+N}{2}}$$

Here $\Gamma(\lambda)$ is the well-known Gamma function given by

$$\Gamma(\lambda) = \int_0^{+\infty} t^{\lambda-1} e^{-t} dt$$

The class of elliptical distributions is a natural generalization of the class of Gaussian distributions. Many properties of Gaussian distributions have analogs for elliptical distributions, but this class is much larger, in particular it includes distributions with heavy tails. For detailed investigation of elliptical distributions, see [1, 32]. It is known that if $E(X)$ exists then $E(X) = \mu$. One important property of elliptical distribution X is the connection between covariance matrix of the vector X and the matrix Λ. Namely, if covariance matrix exists one has

$$\sigma_{i,j} = \text{Cov}(X_i, X_j) = C \cdot \lambda_{i,j} \tag{2.2}$$

where

$$C = \frac{2\pi^{\frac{1}{2}N}}{\Gamma(\frac{1}{2}N)} \int_0^{+\infty} r^{N+1} g(r^2) dr$$

In particular, for Gaussian distribution one has $\text{Cov}(X_i, X_j) = \lambda_{i,j}$. For multivariate Student distribution with ν degree of freedom ($\nu > 2$), one has $\sigma_{i,j} = \nu/(\nu-1)\lambda_{i,j}$.

In this section, we discuss several measures of dependence (similarity) known as measures of association studied by Kruskal [57]. In our opinion, this important paper was not fully appreciated in the literature. Let us consider two-dimensional random vector (X, Y). The classical Pearson correlation is defined by

$$\gamma^P(X, Y) = \frac{\text{Cov}(X, Y)}{\sqrt{\text{Cov}(X, X)} \sqrt{\text{Cov}(Y, Y)}} \tag{2.3}$$

Common interpretation of $\gamma^P(X, Y)$ is a measure of linear dependence between Y and X. There is an old and interesting interpretation of this correlation [45]. Suppose that the structure of X and Y has the form:

$$X = U_1 + U_2 + \ldots + U_m + V_1 + \ldots + V_n$$

$$Y = U_1 + U_2 + \ldots + U_m + W_1 + \ldots + W_n$$

where U, V, W are all mutually uncorrelated with the same variance. Then $\gamma^P(X, Y) = \frac{m}{m+n}$ which is the proportion of common components between X and Y. Pearson correlation is the most popular measure of similarity in network analysis.

From the other side, there is a family of measures of similarity (association) between X and Y on the base of probabilities

$$P\{(X > x_0 \text{ and } Y > y_0) \text{ or } (X < x_0 \text{ and } Y < y_0)\} = P\{(X - x_0)(Y - y_0) > 0\},$$

where x_0, y_0 are some real numbers [57]. These probabilities have a natural interpretation as the proportion of concordance of deviation of X and Y from x_0 and y_0. Different choice of x_0, y_0 leads to different measures of similarity. In this book, we emphasize the sign similarity measure of association. This measure is obtained by the following choice of x_0, y_0: $x_0 = E(X)$, $y_0 = E(Y)$

$$\gamma^{Sg}(X, Y) = P\{(X - E(X))(Y - E(Y)) > 0\} \tag{2.4}$$

Sign similarity is related with Fechner correlation

$$\gamma^{Fh}(X, Y) = 2\gamma^{Sg}(X, Y) - 1 \tag{2.5}$$

Sign similarity measure of association is the proportion of concordance of deviation of X and Y from their expected values.

For $x_0 = \text{Med}X$ and $y_0 = \text{Med}Y$, one has Kruskal correlation

$$\gamma^{Kr}(X, Y) = 2P\{(X - \text{Med}X)(Y - \text{Med}Y) > 0\} - 1 \tag{2.6}$$

Following [57], we notice that $\gamma^{Kr}(X, Y)$ remains unchanged by monotone functional transformations of the coordinates: if instead of X and Y we consider $f(X)$ and $g(Y)$, where f and g are both monotone strictly increasing (decreasing) then $\gamma^{Kr}(X, Y)$ is unchanged.

In order to avoid arbitrariness in the choice of x_0, y_0, one can consider the difference between two independent random vectors (X_1, Y_1), (X_2, Y_2) with the same distribution as (X, Y). This leads to a measure of similarity connected with Kendall-τ correlation. Define

$$\gamma^{Kd}(X, Y) = 2P\{(X_1 - X_2)(Y_1 - Y_2) > 0\} - 1 \tag{2.7}$$

The classical Kendall-τ correlation can be considered as unbiased and consistent estimation of $\gamma^{Kd}(X, Y)$. There is an obvious relation between $\gamma^{Kd}(X, Y)$, $\gamma^{Kr}(X, Y)$ and $\gamma^{Fh}(X, Y)$:

$$\gamma^{Kd}(X, Y) = \gamma^{Kr}(X_1 - X_2, Y_1 - Y_2) = \gamma^{Fh}(X_1 - X_2, Y_1 - Y_2)$$

If we consider three independent random vectors (X_1, Y_1), (X_2, Y_2), (X_3, Y_3) with the same distribution as (X, Y) then one can define the following measure of similarity:

$$\gamma^{Sp}(X, Y) = 6P\{(X_1 - X_2)(Y_1 - Y_3) > 0\} - 3 \tag{2.8}$$

The classical Spearman correlation can be considered as unbiased and consistent estimation of $\gamma^{Sp}(X, Y)$. Both measures of similarity $\gamma^{Kd}(X, Y)$, $\gamma^{Sp}(X, Y)$ also remain unchanged by monotone functional transformations of the coordinates. Note that the considered interpretation (proposed in [57]) of the Kendall-τ correlation and the Spearman correlation is not traditional. In our investigations, it is important to define the Kendall-τ correlation and the Spearman correlation as measures of similarity between two random variables in order to construct true network structures in associated network models.

Different distributions and different measures of similarity generate different random variables networks. The choice of the measure of similarity is crucial for network analysis. Traditionally, Pearson correlation is the most used. This measure is appropriate for Gaussian distribution. However, there is no theoretical justification for the use of Pearson correlation for other type of distributions. In this book, we show that for the class of elliptical distributions the sign similarity is more appropriate for network analysis than Pearson correlation.

2.3 *Relations* Between Different Networks and Network Structures

First we investigate the relations between different measures of similarities for two dimensional elliptical distributions. Let (X, Y) be two dimensional vector with elliptical distribution with density function

$$f(x, y) = \det(A)^{-\frac{1}{2}} g(a^{1,1}(x - \mu_X)^2 + 2a^{1,2}(x - \mu_X)(y - \mu_Y) + a^{2,2}(y - \mu_Y)^2) \tag{2.9}$$

where $A = (a_{i,j})_{i=1,2; j=1,2}$ is positive definite symmetric matrix, and $a^{i,j}$ are the elements of the inverse matrix $A^{-1} = (a^{i,j})_{i=1,2; j=1,2}$, and

$$\int_{-\infty}^{\infty} \int_{-\infty}^{\infty} g(u^2 + v^2) du dv = 1.$$

The density function $f(x, y)$ is symmetric with respect to the vertical line at value (μ_X, μ_Y). It implies that $\text{Med}(X) = \mu_X$, $\text{Med}(Y) = \mu_Y$. If $E(X)$, $E(Y)$ exist then $E(X) = \mu_X = \text{Med}(X)$, $E(Y) = \mu_Y = \text{Med}(Y)$ and Kruskal and Fechner correlations coincide $\gamma^{Kr}(X, Y) = \gamma^{Fh}(X, Y)$. For simplicity of notations sometimes, we will write $\gamma(X, Y) = \gamma$ if it does not lead to some confusions. In the case, when (X, Y) has bi-variate Gaussian distribution, one has [57]:

$$\gamma^{Fh} = \gamma^{Kr} = \gamma^{Kd} = \frac{2}{\pi} \arcsin(\gamma^P), \quad \gamma^{Sg} = \frac{1}{2} + \frac{1}{\pi} \arcsin(\gamma^P) \qquad (2.10)$$

and

$$\gamma^{Sp} = \frac{6}{\pi} \arcsin \frac{\gamma^P}{2} \qquad (2.11)$$

Formulas (2.10) and (2.11) are well known and can be used to get an interpretation of Pearson correlation in the case of Gaussian distribution. For example, $\gamma^P(X, Y) = 0,600$ means that γ^{Sg}, proportion of concordance of deviation of X and Y from $E(X)$ and $E(Y)$, is roughly equal to $0,705$.

In the general case of bi-variate distribution of random vector (X, Y), one can only state that the following inequalities are sharp [57]:

$$-1 + \frac{1}{2}(1 + \gamma^{Kd})^2 \le \gamma^{Sp} \le \frac{1}{2}(1 + 3\gamma^{Kd}) \text{ if } \gamma^{Kd} \le 0$$

$$\frac{1}{2}(-1 + 3\gamma^{Kd}) \le \gamma^{Sp} \le 1 - \frac{1}{2}(1 - \gamma^{Kd})^2 \text{ if } \gamma^{Kd} \ge 0$$

$$\frac{1}{4}(1 + \gamma^{Kr})^2 - 1 \le \gamma^{Kd} \le 1 - \frac{1}{4}(1 - \gamma^{Kr})^2$$

$$\frac{3}{16}(1 + \gamma^{Kr})^3 - 1 \le \gamma^{Sp} \le 1 - \frac{3}{16}(1 - \gamma^{Kr})^3$$

Note that in general case if X and Y are independent then $\gamma^P(X, Y) = \gamma^{Kr}(X, Y) = \gamma^{Kd}(X, Y) = \gamma^{Sp}(X, Y) = 0$ and if in addition $E(X) = \text{Med}(X)$, $E(Y) = \text{Med}(Y)$ then $\gamma^{Fh}(X, Y) = 0$.

For a given positive definite symmetric matrix $A = (a_{i,j})_{i=1,2; j=1,2}$, let us introduce the class $K(A)$ of random vectors (X, Y) with the density function given by (2.9) with arbitrary function g. From general property of elliptical distributions, one has (if all covariances exist):

$$\gamma^P(X, Y) = \frac{\text{Cov}(X, Y)}{\sqrt{\text{Cov}(X, X)}\sqrt{\text{Cov}(Y, Y)}} = \frac{a_{i,j}}{\sqrt{a_{i,i}a_{j,j}}}$$

Therefore, the Pearson correlations $\gamma^P(X, Y)$ are the same for any elliptical vector (X, Y) from the class $K(A)$, such that $E(X^2 + Y^2) < \infty$.

Now we prove that the relations (2.10) between γ^{Fh}, γ^{Kr}, γ^{Kd}, γ^P, γ^{Sg} true for Gaussian vector, remain valid for the case where random vector (X, Y) has bivariate elliptically contoured distribution. For Kendall-τ correlation, it was proved in [65] and [25]. We take a slightly different approach. The main result is

Theorem 2.1 *Let* (X, Y) *be a random vector with density function* (2.9), *and* $E(X^2 + Y^2) < \infty$, *then*

$$\gamma^{Fh} = \gamma^{Kr} = \gamma^{Kd} = \frac{2}{\pi}arcsin(\gamma^P), \quad \gamma^{Sg} = \frac{1}{2} + \frac{1}{\pi}arcsin(\gamma^P) \qquad (2.12)$$

We split the proof on three lemmas.

Lemma 2.1 *Let* (X, Y) *be random vector from the class* $K(A)$. *Then*

$$\gamma^{Sg}(X, Y) = \gamma^{Sg}(X_G, Y_G) \qquad (2.13)$$

where (X_G, Y_G) *is the Gaussian vector from the class* $K(A)$.

Proof Let (X, Y) be a random vector from the class $K(A)$ with the density function (2.9). Without loss of generality, one can suppose $\mu_X = 0$, $\mu_Y = 0$. In this case, one has

$$\gamma^{Sg}(X, Y) = P(X > 0, Y > 0) + P(X < 0, Y < 0).$$

Write

$$P(X > 0, Y > 0) = \det(A)^{-\frac{1}{2}} \int_0^\infty \int_0^\infty g(a^{1,1}x^2 + 2a^{1,2}xy + a^{2,2}y^2)dxdy =$$

$$= \det(A)^{-\frac{1}{2}} \int_0^\infty \int_0^\infty g((x, y)A^{-1}(x, y)')dxdy,$$

where B' means the transpose of the matrix (or vector) B. Matrix A^{-1} is symmetric positive definite. Therefore, there exists matrix $C = (c_{i,j})_{i=1,2; j=1,2}$ such that

$$C'A^{-1}C = \begin{pmatrix} 1 & 0 \\ 0 & 1 \end{pmatrix}.$$

By the change of variables $x = c_{i,i}u + c_{j,i}v$, $y = c_{i,j}u + c_{j,j}v$, that is $(x, y) = (u, v)C'$ one get

$$\det(A)^{-\frac{1}{2}} \int_0^\infty \int_0^\infty g((x, y)A^{-1}(x, y)')dxdy = \det(A)^{-\frac{1}{2}}\det(C)\int_D g(u^2 + v^2)dudv$$

where $D = \{c_{i,i}u + c_{j,i}v > 0, c_{i,j}u + c_{j,j}v > 0\}$. One has

$$(C')^{-1}C^{-1} = (C')^{-1}C'A^{-1}CC^{-1} = A^{-1}$$

It implies $CC' = A$ and $\det(C) = \det(C') = \sqrt{\det(A)}$.

The domain D is a cone (angle) with vertex at $(0, 0)$. In polar coordinates (r, ϕ), it is defined by $0 < r < \infty$, $\phi_1 < \phi < \phi_2$, where ϕ_1, ϕ_2 are defined by the matrix C and does not depend on g. It implies

$$P(X > 0, Y > 0) = \int_D g(u^2 + v^2)dudv = \int_0^\infty \int_{\phi_1}^{\phi_2} g(r^2)rdrd\phi =$$

$$= (\phi_2 - \phi_1)\int_0^\infty g(r^2)rdr = \frac{(\phi_2 - \phi_1)}{2\pi}$$

Therefore, $P(X > 0, Y > 0)$ does not depend on g. In the same way one can prove that $P(X < 0, Y < 0)$ does not depend on g. It implies that $\gamma^{Sg}(X, Y)$ does not depend on g and takes the same value for all distributions from the class $K(\Lambda)$. The Lemma follows.

Now we prove the relation between Pearson correlation and sign similarity for any elliptical bi-variate vector.

Lemma 2.2 *For any bi-variate elliptical vector* (X, Y), *if* $E(X^2 + Y^2) < \infty$ *then*

$$\gamma^{Sg}(X, Y) = \frac{1}{2} + \frac{1}{\pi}arcsin(\gamma^P(X, Y))$$

Proof Let (X, Y) be a random vector with density function (2.9), and (X_G, Y_G) be the Gaussian vector from the same class $K(A)$. One has

$$\gamma^P(X, Y) = \gamma^P(X_G, Y_G), \qquad \gamma^{Sg}(X, Y) = \gamma^{Sg}(X_G, Y_G)$$

Then

$$\gamma^{Sg}(X, Y) = \gamma^{Sg}(X_G, Y_G) = \frac{1}{2} + \frac{1}{\pi}arcsin(\gamma^P(X_G, Y_G)) = \frac{1}{2} + \frac{1}{\pi}arcsin(\gamma^P(X, Y))$$

the lemma is proved.

Next we prove the relations between Pearson, Fechner, Kruskal, Kendall-τ correlations and sign similarity.

Lemma 2.3 *For any bi-variate elliptical vector* (X, Y), *if* $E(X^2 + Y^2) < \infty$ *then*

$$\gamma^{Fh}(X, Y) = \gamma^{Kr}(X, Y) = \gamma^{Kd}(X, Y) = \frac{2}{\pi}arcsin(\gamma^P(X, Y))$$

and

$$\gamma^{Kd}(X, Y) = 2\gamma^{Sg}(X, Y) - 1$$

Proof Let (X_1, Y_1), (X_2, Y_2) be independent random vectors with the same distribution as (X, Y). It can be proved that vector $(X_1 - X_2, Y_1 - Y_2)$ is elliptical too. Therefore,

$$\gamma^{Sg}(X_1 - X_2, Y_1 - Y_2) = \frac{1}{2} + \frac{1}{\pi}\arcsin(\gamma^P(X_1 - X_2, Y_1 - Y_2))$$

One has

$$\text{cov}(X_1 - X_2, Y_1 - Y_2) = 2\text{cov}(X, Y), \quad \text{var}(X_1 - X_2) = 2\text{var}(X), \quad \text{var}(Y_1 - Y_2) = 2\text{var}(Y)$$

It implies

$$\gamma^{Kd}(X, Y) = 2\gamma^{Sg}(X_1 - X_2, Y_1 - Y_2) - 1 = \frac{2}{\pi}\arcsin(\gamma^P(X, Y))$$

The theorem is proved.

Let $K(\Lambda)$ be the class of elliptical distributions defined by the density function (2.1) with fixed matrix Λ. It follows from the statements above that network models and network structures for different random variables networks are related if $X \in K(\Lambda)$. Indeed, if $X \in K(\Lambda)$ then the Fechner correlation network model, the Kruskal correlation network model, and the Kendall correlation network model coincide. The same is true for all network structures in these network models. All these network models are related with Pearson correlation network model by monotone transformation of the weights of edges. The same is true for the sign similarity network model. Therefore, MST and PMFG network structures are the same for the following network models: Fechner correlation network model, Kruskal correlation network model, Kendall correlation network model, Pearson correlation network model, and sign similarity network model. Moreover, threshold graph in one model can be obtained as a threshold graph in another model by an appropriate choice of thresholds. The same is true for maximum cliques and maximum independent sets in the threshold graph. Taking this into account the following question is crucial: are there any differences in network structure identification using different correlations, and if so which correlation is more appropriate? In what follows we show that the use of different correlations generates a different uncertainty in identification and discuss the choice of correlation.

2.4 Partial Correlation Network

Let $X = (X_1, X_2, \ldots, X_N)$ be a random vector. Consider the conditional distribution of the vector (X_i, X_j) when other $X_k, k = 1, \ldots, N; k \neq i, k \neq j$ being fixed. Correlation between X_i, X_j when other $X_k, k = 1, \ldots, N; k \neq i, k \neq j$ being fixed is known as saturated partial correlation and we will denote it by $\gamma_{i,j}^{\text{Par}}$. This partial correlation is used as the measure of similarity in gene expression network analysis [58]. If X has elliptical distribution with the density function (2.1), then there is a connection of partial correlations with the matrix Λ:

$$\gamma_{i,j}^{\text{Par}} = -\frac{\lambda^{i,j}}{\sqrt{\lambda^{i,i} \lambda^{j,j}}},$$

where $\lambda^{i,j}$ are the elements of the inverse matrix Λ^{-1}. Identification of concentration graph by observations in Gaussian partial correlation network is known as the Gaussian Graphical Model Selection Problem [19]. The relations between partial correlations and matrix Λ imply that for all distributions $X \in K(\Lambda)$, the associated network models (V, Γ) coincide. Therefore, all concentration graphs, and other network structures, are the same in partial correlation network for all distributions $X \in K(\Lambda)$.

Chapter 3
Network Structure Identification Algorithms

Abstract In this chapter we state the problem of network structure identification by observations. We discuss in details the problem of threshold graph identification. We show that this problem can be considered as multiple testing problem and describe different multiple testing algorithms for threshold graph identification. We note that existing practice of market graph identification can be considered in the proposed framework. In the same way, we discuss the problem of identification of concentration graph. In addition, we describe some algorithms for MST and PMFG identification. We use numerical simulations to illustrate described algorithms and discuss the results.

3.1 Threshold Graph Identification: Multiple Testing Algorithms

Let (X, γ) be a random variable network. For a given threshold γ_0, the threshold graph in network model (V, Γ) is constructed as follows: the edge between two vertices i and j is included in the threshold graph, iff $\gamma(X_i, X_j) = \gamma_{i,j} > \gamma_0$. We call it *true threshold graph*. In practice $\gamma_{i,j}$ (true network model) are not known. We have only observations from the distribution of the vector X. The *threshold graph identification problem* is to identify the true threshold graph from observations. This problem can be considered as multiple testing problem. Consider the set of individual hypotheses:

$$h_{i,j} : \gamma_{i,j} \leq \gamma_0 \quad \text{vs} \quad k_{i,j} : \gamma_{i,j} > \gamma_0 \ (i, j = 1, \ldots, N; i \neq j). \tag{3.1}$$

We shall assume that tests for the individual hypotheses are available and have the form

$$\varphi_{i,j}(x) = \begin{cases} 0, & T_{i,j}(x) \leq c_{i,j} \\ 1, & T_{i,j}(x) > c_{i,j} \end{cases} \tag{3.2}$$

V. A. Kalyagin et al., *Statistical Analysis of Graph Structures in Random Variable Networks*, SpringerBriefs in Optimization,
https://doi.org/10.1007/978-3-030-60293-2_3

where $T_{i,j}(x)$ are individual tests statistics $i, j = 1, 2, \ldots, N$, $i \neq j$. Define the p-value $p(i, j)$ of the test $\varphi_{i,j}$ as

$$p(i, j) = P_{\gamma_0}(T_{i,j} > t_{i,j}) \tag{3.3}$$

where $t_{i,j}$ is the observed value of $T_{i,j}$, i.e., $t_{i,j} = T_{i,j}(x)$, and the probability P_{γ_0} is calculated by distribution of the statistic $T_{i,j}$ under condition $\gamma_{i,j} = \gamma_0$. In what follows we describe some threshold graph identification algorithms associated with multiple testing procedures.

3.1.1   Simultaneous Inference

Calculate p-value $p(i, j)$ of the test $\varphi_{i,j}$, $i, j = 1, \ldots, N$; $i \neq j$. Fix significance level $\alpha_{i,j}$ for each individual hypothesis $h_{i,j}$.

- If $p(i, j) < \alpha_{i,j}$ then the hypothesis $h_{i,j}$ is rejected, i.e., the edge (i, j) is included in the threshold graph.
- If $p(i, j) \geq \alpha_{i,j}$ then the hypothesis $h_{i,j}$ is accepted, i.e., the edge (i, j) is not included in the threshold graph.

Note that the popular Bonferroni procedure is a particular case of the simultaneous inference if one put $\alpha_{i,j} = \alpha/M$, where M is the number of individual hypothesis $M = N(N - 1)/2$, α is number from the interval $(0, 1)$. It is known that for this procedure the probability of at least one false rejection (false edge inclusion) is bounded by α [62].

3.1.2   Holm Step Down Procedure

Calculate p-value $p(i, j)$ of the test $\varphi_{i,j}$, $i, j = 1, \ldots, N$, $i \neq j$. Fix $\alpha \in (0, 1)$. The step down procedure consists of at most M steps, $M = N(N - 1)/2$. At each step either one individual hypothesis $h_{i,j}$ is rejected or all remaining hypotheses are accepted. Holm step down procedure is constructed as follows:

- Step 1: If

$$\min p(i, j) \geq \frac{\alpha}{M}$$

then accept all hypotheses $h_{i,j}$, $i, j = 1, 2, \ldots, N$ (all vertices in the threshold graph are isolated, there are no edges), else if

$$\min p(i, j) = p(i_1, j_1) < \frac{\alpha}{M}$$

then reject hypothesis h_{i_1,j_1} (include the edge (i_1, j_1) in the threshold graph) and go to step 2.

- ...
- Step K: Let $I = \{(i_1, j_1), (i_2, j_2), \ldots, (i_{K-1}, j_{K-1})\}$ be the set of indices of previously rejected hypotheses (set of edges already included in the threshold graph). If

$$\min_{(i,j)\notin I} p(i, j) \geq \frac{\alpha}{M - K + 1}$$

then accept all hypotheses $h_{i,j}$, $(i, j) \notin I$ (only edges from I are included in the threshold graph),
    else if

$$\min_{(i,j)\notin I} p(i, j) = p(i_K, j_K) < \frac{\alpha}{M - K + 1}$$

then reject hypothesis h_{i_K,j_K} (include edge (i_K, j_k) in the threshold graph) and go to step (K+1).

- ...
- Step M: Let $I = \{(i_1, j_1), \ldots, (i_{M-1}, j_{M-1})\}$ be the set of indices of previously rejected hypotheses. Let $(i_M, j_M) \notin I$. If $p(i_M, j_M) \geq \alpha$ then accept the hypothesis h_{i_M,j_M} (the edge (i_M, j_M) is not included in the threshold graph), else reject hypothesis h_{i_M,j_M} (reject all hypotheses, i.e., the threshold graph is a complete graph).

It is known that for this procedure the probability of at least one false rejection (false edge inclusion) is bounded by α [37]. This procedure is more powerful than Bonferroni procedure [62], i.e., the threshold graph constructed by Bonferroni procedure is a subgraph of the threshold graph constructed by Holm procedure.

3.1.3   Hochberg Step-up Procedure

Calculate p-value $p(i, j)$ of the test $\varphi_{i,j}$, $i, j = 1, \ldots, N$, $i \neq j$. Fix $\alpha \in (0, 1)$. The step-up procedure consists of at most M steps, $M = N(N - 1)/2$. At each step either one individual hypothesis $h_{i,j}$ is accepted or all remaining hypotheses are rejected. Hochberg step-up procedure is constructed as follows:

- Step 1: If

$$\max p(i, j) < \alpha$$

then reject all hypotheses $h_{i,j}$, $i, j = 1, 2, \ldots, N$ (threshold graph is a complete graph), else if

$$\max p(i, j) = p(i_1, j_1) \geq \alpha$$

then accept the hypothesis h_{i_1, j_1} (the edge (i_1, j_1) is not included in the threshold graph) and go to step 2.

- ...
- Step K: Let $I = \{(i_1, j_1), (i_2, j_2), \ldots, (i_{K-1}, j_{K-1})\}$ be the set of indices of previously accepted hypotheses (set of edges already not included in the threshold graph). If

$$\max_{(i,j)\notin I} p(i, j) < \frac{\alpha}{K}$$

then reject all hypotheses $h_{i,j}$, $(i, j) \notin I$ (only edges not from I are included in the threshold graph),

else if

$$\max_{(i,j)\notin I} p(i, j) = p(i_K, j_K) \geq \frac{\alpha}{K}$$

then accept the hypothesis h_{i_K, j_K} (the edge (i_K, j_k) is not included in the threshold graph) and go to step (K+1).

- ...
- Step M: Let $I = \{(i_1, j_1), \ldots, (i_{M-1}, j_{M-1})\}$ be the set of indices of previously accepted hypotheses. Let $(i_M, j_M) \notin I$. If

$$p(i_M, j_M) < \frac{\alpha}{M}$$

then reject the hypothesis h_{i_M, j_M} (the edge (i_M, j_M) is included in the threshold graph,

else accept hypothesis h_{i_M, j_M} (accept all hypotheses, i.e., the threshold graph has isolated vertices).

It is known that this procedure is more powerful than the Holm procedure [35], i.e., the threshold graph constructed by the Holm procedure is a subgraph of the threshold graph constructed by Hochberg procedure. Control of the probability of at least one false rejection (false edge inclusion) is proved under assumption of positive dependence of the components of the vector X [79].

3.1.4 Benjamini-Hochberg Procedure

Calculate p-values $p(i, j)$ of the tests $\varphi_{i,j}$, $i, j = 1, \ldots, N$, $i \neq j$. Fix $\alpha \in (0, 1)$. Benjamini-Hochberg step-up procedure is constructed as follows:

- Step 1: If

$$\max p(i, j) \leq \frac{M\alpha}{M} = \alpha$$

then reject all hypotheses $h_{i,j}$, $i, j = 1, 2, \ldots, N$ (threshold graph is a complete graph), else if

$$\max p(i, j) = p(i_1, j_1) > \alpha$$

then accept the hypothesis h_{i_1, j_1} (the edge (i_1, j_1) is not included in the threshold graph) and go to step 2.

- ...
- Step K: Let $I = \{(i_1, j_1), (i_2, j_2), \ldots, (i_{K-1}, j_{K-1})\}$ be the set of indices of previously accepted hypotheses (set of edges already not included in the threshold graph). If

$$\max_{(i,j) \notin I} p(i, j) \leq \frac{(M - K + 1)\alpha}{M}$$

then reject all hypotheses $h_{i,j}$, $(i, j) \notin I$ (only edges not from I are included in the threshold graph), else if

$$\max_{(i,j) \notin I} p(i, j) = p(i_K, j_K) > \frac{(M - K + 1)\alpha}{M}$$

then accept the hypothesis h_{i_K, j_K} (the edge (i_K, j_K) is not included in the threshold graph) and go to step (K+1).

- ...
- Step M: Let $I = \{(i_1, j_1), \ldots, (i_{M-1}, j_{M-1})\}$ be the set of indices of previously accepted hypotheses. Let $(i_M, j_M) \notin I$. If

$$p(i_M, j_M) \leq \frac{\alpha}{M}$$

then reject the hypothesis h_{i_M, j_M} (the edge (i_M, j_M) is included in the threshold graph), else accept hypothesis h_{i_M, j_M} (accept all hypotheses, i.e., the threshold graph has isolated vertices).

It is known that in some cases the expectation of proportion of false rejections for Benjamini-Hochberg procedure is bounded by α [2]. Moreover, Benjamini-Hochberg procedure is more powerful than Hochberg procedure, i.e., the threshold graph constructed by Hochberg procedure is a subgraph of the threshold graph constructed by Benjamini-Hochberg procedure.

3.1.5   Individual Test Statistics

Let $x(t)$ be a sample of the size n from distribution of the random vector X:

$$x(t) = (x_1(t), x_2(t), \ldots, x_N(t)), \ t = 1, 2, \ldots, n$$

In this section, we describe the individual test statistics and calculate their p-values for different random variable networks.

Pearson correlation network   For Pearson correlation network model, individual hypotheses (3.1) have the form:

$$h_{i,j} : \gamma_{i,j}^P \leq \gamma_0^P.$$

For Gaussian distribution of the vector X, statistics $T_{i,j}$ for individual hypotheses testing can be taken from [1]:

$$T_{i,j}^P = \sqrt{n} \left(\frac{1}{2} \ln \left(\frac{1 + r_{i,j}}{1 - r_{i,j}} \right) - \frac{1}{2} \ln \left(\frac{1 + \gamma_0^P}{1 - \gamma_0^P} \right) \right)$$

where

$$r_{i,j} = \frac{\sum_t (x_i(t) - \bar{x}_i)(x_j(t) - \bar{x}_j)}{\sqrt{\sum_t (x_i(t) - \bar{x}_i)^2 \sum_t (x_j(t) - \bar{x}_j)^2}}$$

and $\bar{x}_i = (1/n) \sum_t x_i(t)$. Note that the distribution of $r_{i,j}$ is known (for Gaussian bi-dimensional vector) and can be found in [1]. According to [1], asymptotic distribution of $T_{i,j}^P$ for $\gamma_{i,j}^P = \gamma_0^P$ is the standard Gaussian distribution. Therefore, p-values for simultaneous inference, Holm, Hochberg, and Benjamini-Hochberg procedures can be calculated from equations

$$p^P(i, j) = 1 - \Phi(t_{i,j}^P)$$

where $\Phi(x)$ is the cumulative distribution function for the standard Gaussian distribution, and $t_{i,j}^P$ is the observed value of the statistic $T_{i,j}^P$.

For elliptical distribution of the vector X, the following statistics has asymptotically standard Gaussian distribution [1]

$$\sqrt{\frac{n}{1 + \hat{k}}} T_{i,j}^P$$

where $\hat{k}$ is a sample analog of kurtosis and

$$1 + \hat{k} = \frac{1}{N(N+2)} \frac{1}{n} \sum_{t=1}^{n} \left[(x(t) - \bar{x})' S^{-1} (x(t) - \bar{x}) \right]^2$$

and S is the sample covariance matrix $S = (s_{i,j})$

$$s_{i,j} = \frac{1}{n} \sum_{t=1}^{n} (x_i(t) - \bar{x}_i)(x_j(t) - \bar{x}_j)$$

In this case, the p-values for simultaneous inference, Holm, Hochberg, and Benjamini-Hochberg procedures can be calculated from equations

$$p^P(i, j) = 1 - \Phi \left(\sqrt{\frac{n}{1 + \hat{k}}} t_{i,j}^P \right)$$

Sign similarity network For sign similarity network, individual hypotheses have the form:

$$h_{i,j} : \gamma_{i,j}^{Sg} \leq \gamma_0^{Sg}.$$

Suppose the vector $\mu = E(X)$ is known. Define

$$I_{i,j}^{Sg}(t) = \begin{cases} 1, & (x_i(t) - \mu_i)(x_j(t) - \mu_j) \geq 0 \\ 0, & (x_i(t) - \mu_i)(x_j(t) - \mu_j) < 0 \end{cases}$$

$$T_{i,j}^{sg} = \sum_{t=1}^{n} I_{i,j}^{Sg}(t)$$

Statistic $T_{i,j}^{Sg}$ for $\gamma_{i,j}^{Sg} = \gamma_0^{Sg}$ has a binomial distribution with parameters (n, γ_0^{Sg}). Therefore, p-values for simultaneous inference, Holm, Hochberg, and Benjamini-Hochberg procedures can be calculated from the equations

$$p^{Sg}(i, j) = 1 - F(t_{i,j}^{Sg}; n, \gamma_0^{Sg})$$

where $F(t; n, \gamma_0^{Sg})$ is the cumulative distribution function for binomial distribution with parameters (n, γ_0^{Sg}). When μ is unknown, one can use the statistic $T_{i,j}^{Sg}$ with

$$I_{i,j}^{Sg}(t) = \begin{cases} 1, & (x_i(t) - \bar{x}_i)(x_j(t) - \bar{x}_j) \geq 0 \\ 0, & (x_i(t) - \bar{x}_i)(x_j(t) - \bar{x}_j) < 0 \end{cases}$$

Note that for elliptical distributions, the following statistic

$$T_{\text{norm}(i,j)}^{Sg} = \frac{T_{i,j}^{Sg} - n\gamma_0^{Sg}}{\sqrt{n\gamma_0^{Sg}(1 - \gamma_0^{Sg})}}$$

has asymptotically standard Gaussian distribution. This can be used for estimation of p-values for a large n, i.e., p-values can be calculated from the equation

$$p^{Sg}(i, j) = 1 - \Phi(t_{\text{norm}(i,j)}^{Sg})$$

Fechner correlation network   For Fechner correlation, one can use the following estimation:

$$\hat{\gamma}_{i,j}^{Fh} = \frac{1}{n} \sum_{t=1}^{n} I_{i,j}^{Fh}(t)$$

where

$$I_{i,j}^{Fh}(t) = \begin{cases} 1, & (x_i(t) - \overline{x_i})(x_j(t) - \overline{x_j}) \geq 0 \\ -1, & (x_i(t) - \overline{x_i})(x_j(t) - \overline{x_j}) < 0 \end{cases}$$

One has $I_{i,j}^{Fh}(t) = 2I_{i,j}^{Sg}(t) - 1$ and for the p-values of tests for Fechner correlations one can use statistics $T_{i,j}^{Sg}$ with $\gamma_0^{Sg} = (1 + \gamma_0^{Fh})/2$.

Kruskal correlation network   For Kruskal correlation, one can use the following estimation:

$$\hat{\gamma}_{i,j}^{Kr} = \frac{1}{n} \sum_{t=1}^{n} I_{i,j}^{Kr}(t)$$

where

$$I_{i,j}^{Kr}(t) = \begin{cases} 1, & (x_i(t) - \text{med}(x_i))(x_j(t) - \text{med}(x_j)) \geq 0 \\ -1, & (x_i(t) - \text{med}(x_i))(x_j(t) - \text{med}(x_j)) < 0 \end{cases}$$

and $\text{med}(x_i)$ is the sample median of observations $x_i(t)$, $t = 1, 2, \ldots, n$. For elliptical distributions, the following statistic

$$T_{\text{norm}(i,j)}^{Kr} = \sqrt{n} \frac{\hat{\gamma}_{i,j}^{Kr} - \gamma_0^{Kr}}{\sqrt{1 - (\gamma_0^{Kr})^2}}$$

has asymptotically standard Gaussian distribution [3, 57]. Therefore, p-values for Kruskal correlation tests can be calculated from the equations

$$p^{Kr}(i, j) = 1 - \Phi(t^{Kr}_{\text{norm}(i,j)})$$

Kendall correlation network For Kendall correlation, one can use the following estimation:

$$\hat{\gamma}^{Kd}_{i,j} = \frac{1}{n(n-1)} \sum_{t=1}^{n} \sum_{\substack{s=1 \\ s \neq t}}^{n} I^{Kd}_{i,j}(t, s)$$

where

$$I^{Kd}_{i, j}(t, s) = \begin{cases} 1, & (x_i(t) - x_i(s))(x_j(t) - x_j(s)) \geq 0 \\ -1, & (x_i(t) - x_i(s))(x_j(t) - x_j(s)) < 0 \end{cases}$$

For elliptical distributions, the following statistic has asymptotically standard Gaussian distribution [36, 57]

$$T^{Kd}_{\text{norm}(i,j)} = \frac{\sqrt{n}(\hat{\gamma}^{Kd}_{i,j} - \gamma^{Kd}_0)}{4\sqrt{\hat{P}_{cc} - \hat{P}^2_c}},$$

where $\hat{P}_{cc}$ is the estimation of the probability that for any three pairs of observations, the second and third are concordant with the first (concordance with respect to the order), $\hat{P}_c = (\hat{\gamma}^{Kd}_{i,j} + 1)/2$. Therefore, p-values for Kendall correlation tests can be calculated from the equations

$$p^{Kd}(i, j) = 1 - \Phi(t^{Kd}_{\text{norm}(i,j)})$$

Spearman correlation network For Spearman correlation, one can use the following estimation:

$$\hat{\gamma}^{Sp}_{i,j} = \frac{3}{n(n-1)(n-2)} \sum_{t=1}^{n} \sum_{\substack{s=1 \\ s \neq t}}^{n} \sum_{\substack{l=1 \\ l \neq t \\ l \neq s}}^{n} I^{Sp}_{i,j}(t, s, l)$$

where

$$I^{Sp}_{i,j}(t, s, l) = \begin{cases} 1, & (x_i(t) - x_i(s))(x_j(t) - x_j(l)) \geq 0 \\ -1, & (x_i(t) - x_i(s))(x_j(t) - x_j(l)) < 0 \end{cases}$$

For elliptical distributions, the following statistic

$$T^{Sp}_{norm(i,j)} = \sqrt{n-1}(\hat{\gamma}^{Sp}_{i,j} - \gamma^{Sp}_0)$$

has asymptotically standard Gaussian distribution [57]. Therefore, p-values for Spearman correlation tests can be calculated from the equations

$$p^{Sp}(i,j) = 1 - \Phi(t^{Sp}_{norm(i,j)})$$

Partial correlation network Let X has elliptical distribution and $\lambda^{i,j}$ be the elements of the matrix Λ^{-1}. Denote by $\gamma^{Par}_{i,j}$ the partial correlation between X_i, X_j for fixed $X_k, k = 1, \ldots, N; k \neq i, j$. One has [59]

$$\gamma^{Par}_{i,j} = -\frac{\lambda^{i,j}}{\sqrt{\lambda^{i,i}\lambda^{j,j}}}$$

Define the statistics

$$T^{Par}_{i,j} = \sqrt{n}\left(\frac{1}{2}\ln\left(\frac{1+r^{i,j}}{1-r^{i,j}}\right) - \frac{1}{2}\ln\left(\frac{1+\gamma^{Par}_0}{1-\gamma^{Par}_0}\right)\right)$$

where

$$r^{i,j} = -\frac{s^{i,j}}{\sqrt{s^{i,i}s^{j,j}}}$$

and $s^{i,j}$ are the elements of the matrix S^{-1} (inverse to the sample covariance matrix). The following statistics

$$\sqrt{\frac{n}{1+\hat{k}}}T^{Par}_{i,j}$$

have asymptotically standard Gaussian distribution, where $\hat{k}$ is a sample analog of kurtosis and

$$1 + \hat{k} = \frac{1}{N(N+2)}\frac{1}{n}\sum_{t=1}^{n}\left[(x(t) - \overline{x})'S^{-1}(x(t) - \overline{x})\right]^2$$

In this case, the p-values for simultaneous inference, Holm, Hochberg, and Benjamini-Hochberg procedures can be calculated from equations

$$p^{Par}(i,j) = 1 - \Phi(\sqrt{\frac{n}{1+\hat{k}}}t^{Par}_{i,j})$$

Remark The most popular threshold graph is the market graph for the stock market network [6]. For the market graph calculation, the most used measure of similarity is Pearson correlation; algorithm of identification is simultaneous inference with the following individual tests:

$$r_{i,j} > \gamma_0^P, \quad i, j = 1, 2, \ldots, N, \quad i \neq j$$

Since the asymptotic distribution of $r_{i,j}$ is normal with expectation equal to γ_0^P [1], then such test corresponds to the choice of individual significance level $\alpha = 0, 5$. The paper [82] deals with market graph calculation for Spearman correlation network. Algorithm of identification is simultaneous inference with the following individual tests

$$\hat{\gamma}_{i,j}^{Sp} > \gamma_0^{Sp}, \quad i, j = 1, 2, \ldots, N, \quad i \neq j$$

It is interesting to investigate general statistical properties of described procedures for different networks and general choice of significance level.

3.2 Concentration Graph Identification

Let (X, γ) be a random variable network. The concentration graph in a network model (V, Γ) is constructed as follows: the edge between two vertices i and j is included in the concentration graph, iff $\gamma(X_i, X_j) = \gamma_{i,j} \neq 0$. We call it *true concentration graph*. *Concentration graph identification problem* is to identify the true concentration graph from observations. This problem can be considered as multiple testing problem too. Consider the set of individual hypotheses:

$$h_{i,j} : \gamma_{i,j} = 0 \quad \text{vs} \quad k_{i,j} : \gamma_{i,j} \neq 0, \quad (i, j = 1, \ldots, N; i \neq j). \tag{3.4}$$

We shall assume that tests for the individual hypotheses are available and have the symmetric form

$$\varphi_{i,j}(x) = \begin{cases} 0, & |T_{i,j}(x)| \leq c_{i,j} \\ 1, & |T_{i,j}(x)| > c_{i,j} \end{cases} \tag{3.5}$$

where $T_{i,j}(x)$ are individual tests statistics $i, j = 1, 2, \ldots, N, \; i \neq j$. Define the p-value $p(i, j)$ of the test $\varphi_{i,j}$ as

$$p(i, j) = P_0(|T_{i,j}| > |t_{i,j}|) \tag{3.6}$$

where $t_{i,j}$ is the observed value of $T_{i,j}$, i.e., $t_{i,j} = T_{i,j}(x)$, and the probability P_0 is calculated by distribution of the statistic $T_{i,j}$ under condition $\gamma_{i,j} = 0$. One can use the same algorithms (simultaneous inference, Holm, Hochberg, Benjamini-Hochberg) for concentration graph identification as for threshold graph identification.

For the concentration graph identification in Pearson, Fechner, Kruskal, Kendall, Spearman, and partial correlation networks one can use the same statistics as for threshold graph identification for $\gamma_0 = 0$ with the p-values calculated by

$$p(i, j) = 2[1 - \Phi(|t_{i,j}|)]$$

where $t_{i,j}$ is the observed value of associated statistics.

Remark Partial correlation is a popular measure of similarity in gene expression network analysis. Identification of the concentration graph is well studied for the Gaussian distribution of the vector X and in this case the identification problem is called Gaussian Graphical Model Selection problem (GGMS). Different algorithms are known for the solution of GGMS problem [19, 20].

3.3   Maximum Spanning Tree Identification

Let (X, γ) be a random variable network, and (V, Γ) associated network model. The *true maximum spanning tree* in the complete weighted graph (V, Γ) can be constructed by any appropriate algorithm, such as Kruskal algorithm, Prim algorithm, and others [29]. In what follows we will use the Kruskal algorithm.

Kruskal algorithms   To construct the MST, a list of edges is sorted in descending order according to the weight and following the ordered list an edge is added to the MST if and only if it does not create a cycle. In general, MST constructed by Kruskal algorithm is not unique. To avoid complications, we will consider the case where all weights of edges are different. In this case, MST is unique.

In practice $\gamma_{i,j}$ (true network model) are not known. We have only observations from the distribution of the vector X. The *maximum spanning tree identification problem* is to identify the MST from observations. To solve this problem one can use $\hat{\gamma}_{i,j}$, estimation of $\gamma_{i,j}$, to construct a sample network model, and then use any algorithm of maximum spanning tree construction (e.g., Kruskal algorithm). In particular, one can use the following estimations for different networks

Sample Pearson correlation

$$\hat{\gamma}_{i,j}^P = r_{i,j} = \frac{\sum_t (x_i(t) - \overline{x}_i)(x_j(t) - \overline{x}_j)}{\sqrt{\sum_t (x_i(t) - \overline{x}_i)^2 \sum_t (x_j(t) - \overline{x}_j)^2}}$$

Sample sign similarity

$$\hat{\gamma}_{i,j}^{Sg} = \frac{1}{n} T_{i,j}^{Sg} = \frac{1}{n} \sum_{t=1}^{n} I_{i,j}^{Sg}(t)$$

Sample Fechner similarity

$$\hat{\gamma}_{i,j}^{Fh} = \frac{1}{n} \sum_{t=1}^{n} I_{i,j}^{Fh}(t)$$

Sample Kruskal correlation

$$\hat{\gamma}_{i,j}^{Kr} = \frac{1}{n} \sum_{t=1}^{n} I_{i,j}^{Kr}(t)$$

Sample Kendall correlation

$$\hat{\gamma}_{i,j}^{Kd} = \frac{1}{n(n-1)} \sum_{t=1}^{n} \sum_{\substack{s=1 \\ s \neq t}}^{n} I_{i,j}^{Kd}(t, s)$$

Sample Spearman correlation

$$\hat{\gamma}_{i,j}^{Sp} = \frac{3}{n(n-1)(n-2)} \sum_{t=1}^{n} \sum_{\substack{s=1 \\ s \neq t}}^{n} \sum_{\substack{l=1 \\ l \neq t \\ l \neq s}}^{n} I_{i,j}^{Sp}(t, s, l)$$

Sample partial correlation

$$\hat{\gamma}_{i,j}^{Par} = r^{i,j} = -\frac{s^{i,j}}{\sqrt{s^{i,i} s^{j,j}}}$$

Here $s^{i,j}$ are the elements of inverse sample covariance matrix S^{-1}.

Remark MST is widely used in market network analysis. Number of publications is growing [67]. However, the question of uncertainty of MST identification is not well studied.

3.4 Example of MST Identification

We take the example of the Sect. 2.1. Consider the class of elliptical distributions of the class $K(\Lambda)$, where matrix Λ is defined in the Example 2.1. True MST is given by the Fig. 2.1. True MST is the same for all distributions from the class $K(\Lambda)$. By Cayley formula [13] total number of possible MST is 10^8. To reduce the number of variants, we introduce the following simple topological characteristic of MST: vector of degrees of vertices of MST ordered in ascending order. With respect to the characteristic, there are 22 possible topologically different MST, which are listed below:

$\{(1, 1, 1, 1, 1, 1, 1, 1, 1, 9),\ (1, 1, 1, 1, 1, 1, 1, 1, 2, 8),\ (1, 1, 1, 1, 1, 1, 1, 1, 3, 7),$
$(1, 1, 1, 1, 1, 1, 1, 1, 4, 6),\ (1, 1, 1, 1, 1, 1, 1, 1, 5, 5),\ (1, 1, 1, 1, 1, 1, 1, 2, 2, 7),$
$(1, 1, 1, 1, 1, 1, 1, 2, 3, 6),\ (1, 1, 1, 1, 1, 1, 1, 2, 4, 5),\ (1, 1, 1, 1, 1, 1, 1, 3, 3, 5),$
$(1, 1, 1, 1, 1, 1, 1, 3, 4, 4),\ (1, 1, 1, 1, 1, 1, 2, 2, 2, 6),\ (1, 1, 1, 1, 1, 1, 2, 2, 3, 5),$
$(1, 1, 1, 1, 1, 1, 2, 2, 4, 4),\ (1, 1, 1, 1, 1, 1, 2, 3, 3, 4),\ (1, 1, 1, 1, 1, 1, 3, 3, 3, 3),$
$(1, 1, 1, 1, 1, 2, 2, 2, 2, 5),\ (1, 1, 1, 1, 1, 2, 2, 2, 3, 4),\ (1, 1, 1, 1, 1, 2, 2, 3, 3, 3),$
$(1, 1, 1, 1, 2, 2, 2, 2, 2, 4),\ (1, 1, 1, 1, 2, 2, 2, 2, 3, 3),\ (1, 1, 1, 2, 2, 2, 2, 2, 2, 3),$
$(1, 1, 2, 2, 2, 2, 2, 2, 2, 2)\}$

True MST corresponds to the following vector $(1, 1, 1, 1, 1, 1, 1, 1, 5, 5)$. We conduct the following experiments:

- For a given distribution of the vector X from $K(\Lambda)$, generate sample of the size n from X
- Calculate the estimations $\hat{\gamma}_{i,j}$ of the true edge weights in the network
- Apply Kruskal algorithm to construct MST for the sample network $(V, \hat{\Gamma})$, $\hat{\Gamma} = (\hat{\gamma}_{i,j})$
- Calculate the topological characteristic of obtained MST
- Repeat the experiment S times, and calculate the frequencies of appearance of each topological characteristic

We chose two distributions from the class $K(\Lambda)$: Gaussian distribution and Student distribution with 3 degrees of freedom. Sample sizes are going from $n = 5$ to $n = 50{,}000$. Number of replications is $S = 1000$. The results are presented in the Tables 3.1, 3.2, 3.3, and 3.4. Tables 3.1 and 3.2 present the results for Pearson correlation network. One can see that for Gaussian distribution and $n = 5, 10, 20$, the true MST almost not appear, most popular are MST with the following topological characteristics $(1, 1, 1, 1, 2, 2, 2, 2, 3, 3)$, $(1, 1, 1, 1, 1, 2, 2, 2, 3, 4)$. The situation is much worse for Student distribution. Note that for a small number of observations the hubs of the true MST (vertices 9 and 10) are not identified. The true MST structure starts to be identified only from $n = 50{,}000$. Tables 3.3 and 3.4 present the results for sign similarity network. One can see that in this case the picture is much more stable with respect to distribution than for Pearson correlation network. As we

Table 3.1 Observed frequencies of degree vectors for 1000 simulations. *Pearson correlation network, normal distribution*

Degree vec./no. of observations	5	10	20	100	1000	10,000	50,000
(1, 1, 1, 1, 1, 1, 1, 1, 1, 9)	0	0	0	2	0	0	0
(1, 1, 1, 1, 1, 1, 1, 1, 2, 8)	0	0	1	35	47	0	0
(1, 1, 1, 1, 1, 1, 1, 1, 3, 7)	0	0	1	72	164	28	0
(1, 1, 1, 1, 1, 1, 1, 1, 4, 6)	0	0	2	128	339	430	378
(1, 1, 1, 1, 1, 1, 1, 1, 5, 5)	**0**	**0**	**5**	**84**	**255**	**540**	**622**
(1, 1, 1, 1, 1, 1, 1, 2, 2, 7)	0	2	16	67	7	0	0
(1, 1, 1, 1, 1, 1, 1, 2, 3, 6)	0	5	38	180	45	0	0
(1, 1, 1, 1, 1, 1, 1, 2, 4, 5)	0	6	48	217	143	2	0
(1, 1, 1, 1, 1, 1, 1, 3, 3, 5)	0	9	25	18	0	0	0
(1, 1, 1, 1, 1, 1, 1, 3, 4, 4)	2	8	27	17	0	0	0
(1, 1, 1, 1, 1, 1, 2, 2, 2, 6)	0	13	32	35	0	0	0
(1, 1, 1, 1, 1, 1, 2, 2, 3, 5)	1	61	125	67	0	0	0
(1, 1, 1, 1, 1, 1, 2, 2, 4, 4)	4	39	84	44	0	0	0
(1, 1, 1, 1, 1, 1, 2, 3, 3, 4)	16	107	119	14	0	0	0
(1, 1, 1, 1, 1, 1, 3, 3, 3, 3)	5	8	9	0	0	0	0
(1, 1, 1, 1, 1, 2, 2, 2, 2, 5)	3	40	55	7	0	0	0
(1, 1, 1, 1, 1, 2, 2, 2, 3, 4)	107	236	211	12	0	0	0
(1, 1, 1, 1, 1, 2, 2, 3, 3, 3)	171	166	92	1	0	0	0
(1, 1, 1, 1, 2, 2, 2, 2, 2, 4)	61	70	42	0	0	0	0
(1, 1, 1, 1, 2, 2, 2, 2, 3, 3)	410	180	62	0	0	0	0
(1, 1, 1, 2, 2, 2, 2, 2, 2, 3)	210	49	6	0	0	0	0
(1, 1, 2, 2, 2, 2, 2, 2, 2, 2)	10	1	0	0	0	0	0

will show later this fact has a strong theoretical justification. At the same time MST identification in sign similarity network is more correct for Student distribution that in Pearson correlation network.

Table 3.2 Observed frequencies of degree vectors for 1000 simulations. *Pearson correlation network, Student distribution with 3 degree of freedom*

Degree vec./no. of observations	5	10	20	100	1000	10,000	50,000
(1, 1, 1, 1, 1, 1, 1, 1, 1, 9)	0	0	0	1	2	1	0
(1, 1, 1, 1, 1, 1, 1, 1, 2, 8)	0	0	0	8	35	36	32
(1, 1, 1, 1, 1, 1, 1, 1, 3, 7)	0	0	0	9	85	156	116
(1, 1, 1, 1, 1, 1, 1, 1, 4, 6)	0	1	1	24	148	279	351
(1, 1, 1, 1, 1, 1, 1, 1, 5, 5)	**0**	**0**	**0**	**13**	**95**	**210**	**318**
(1, 1, 1, 1, 1, 1, 1, 2, 2, 7)	0	0	5	30	44	20	7
(1, 1, 1, 1, 1, 1, 1, 2, 3, 6)	0	1	8	97	147	89	37
(1, 1, 1, 1, 1, 1, 1, 2, 4, 5)	0	2	21	96	226	179	116
(1, 1, 1, 1, 1, 1, 1, 3, 3, 5)	0	1	12	36	20	7	0
(1, 1, 1, 1, 1, 1, 1, 3, 4, 4)	0	3	11	31	19	5	1
(1, 1, 1, 1, 1, 1, 2, 2, 2, 6)	0	4	19	56	35	1	4
(1, 1, 1, 1, 1, 1, 2, 2, 3, 5)	1	38	87	170	54	6	5
(1, 1, 1, 1, 1, 1, 2, 2, 4, 4)	3	36	50	91	37	4	4
(1, 1, 1, 1, 1, 1, 2, 3, 3, 4)	11	58	106	94	16	2	5
(1, 1, 1, 1, 1, 1, 3, 3, 3, 3)	3	13	8	3	0	0	1
(1, 1, 1, 1, 1, 2, 2, 2, 2, 5)	5	21	50	39	7	0	0
(1, 1, 1, 1, 1, 2, 2, 2, 3, 4)	82	217	241	113	20	2	0
(1, 1, 1, 1, 1, 2, 2, 3, 3, 3)	121	146	116	43	5	1	0
(1, 1, 1, 1, 2, 2, 2, 2, 2, 4)	47	84	67	15	4	1	0
(1, 1, 1, 1, 2, 2, 2, 2, 3, 3)	406	278	151	26	1	1	3
(1, 1, 1, 2, 2, 2, 2, 2, 2, 3)	285	94	43	5	0	0	0
(1, 1, 2, 2, 2, 2, 2, 2, 2, 2)	36	3	4	0	0	0	0

Table 3.3 Observed frequencies of degree vectors for 1000 simulations. *Sign similarity network, normal distribution*

Degree vec./no. of observations	5	10	20	100	1000	10,000	50,000
(1, 1, 1, 1, 1, 1, 1, 1, 1, 9)	25	3	0	1	7	0	0
(1, 1, 1, 1, 1, 1, 1, 1, 2, 8)	63	11	8	17	80	31	5
(1, 1, 1, 1, 1, 1, 1, 1, 3, 7)	53	14	4	17	161	188	104
(1, 1, 1, 1, 1, 1, 1, 1, 4, 6)	61	12	6	27	225	370	431
(1, 1, 1, 1, 1, 1, 1, 1, 5, 5)	**35**	**6**	**1**	**14**	**123**	**307**	**456**
(1, 1, 1, 1, 1, 1, 1, 2, 2, 7)	60	33	24	54	55	5	0
(1, 1, 1, 1, 1, 1, 1, 2, 3, 6)	108	53	42	106	120	28	0
(1, 1, 1, 1, 1, 1, 1, 2, 4, 5)	109	57	34	106	194	71	4
(1, 1, 1, 1, 1, 1, 1, 3, 3, 5)	44	22	15	28	8	0	0
(1, 1, 1, 1, 1, 1, 1, 3, 4, 4)	31	21	12	24	6	0	0
(1, 1, 1, 1, 1, 1, 2, 2, 2, 6)	38	63	59	65	6	0	0
(1, 1, 1, 1, 1, 1, 2, 2, 3, 5)	106	111	121	157	8	0	0
(1, 1, 1, 1, 1, 1, 2, 2, 4, 4)	40	70	63	88	6	0	0
(1, 1, 1, 1, 1, 1, 2, 3, 3, 4)	81	101	99	78	0	0	0
(1, 1, 1, 1, 1, 1, 3, 3, 3, 3)	0	5	3	6	0	0	0
(1, 1, 1, 1, 1, 2, 2, 2, 2, 5)	17	44	70	41	0	0	0
(1, 1, 1, 1, 1, 2, 2, 2, 3, 4)	63	189	199	116	1	0	0
(1, 1, 1, 1, 1, 2, 2, 3, 3, 3)	37	61	83	27	0	0	0
(1, 1, 1, 1, 2, 2, 2, 2, 2, 4)	8	34	42	10	0	0	0
(1, 1, 1, 1, 2, 2, 2, 2, 3, 3)	17	69	94	18	0	0	0
(1, 1, 1, 2, 2, 2, 2, 2, 2, 3)	4	20	21	0	0	0	0
(1, 1, 2, 2, 2, 2, 2, 2, 2, 2)	0	1	0	0	0	0	0

Table 3.4 Observed frequencies of degree vectors for 1000 simulations. *Sign similarity network, Student distribution with 3 degree of freedom*

Degree vec./no. of observations	5	10	20	100	1000	10,000	50,000
(1, 1, 1, 1, 1, 1, 1, 1, 1, 9)	24	3	2	0	12	0	0
(1, 1, 1, 1, 1, 1, 1, 1, 2, 8)	72	11	5	14	48	32	7
(1, 1, 1, 1, 1, 1, 1, 1, 3, 7)	67	8	3	14	133	165	95
(1, 1, 1, 1, 1, 1, 1, 1, 4, 6)	48	19	3	28	237	418	431
(1, 1, 1, 1, 1, 1, 1, 1, 5, 5)	**38**	**2**	**5**	**16**	**132**	**282**	**461**
(1, 1, 1, 1, 1, 1, 1, 2, 2, 7)	66	28	25	57	46	2	0
(1, 1, 1, 1, 1, 1, 1, 2, 3, 6)	105	53	42	84	125	33	0
(1, 1, 1, 1, 1, 1, 1, 2, 4, 5)	110	48	40	108	219	68	6
(1, 1, 1, 1, 1, 1, 1, 3, 3, 5)	34	22	23	32	5	0	0
(1, 1, 1, 1, 1, 1, 1, 3, 4, 4)	39	17	13	30	7	0	0
(1, 1, 1, 1, 1, 1, 2, 2, 2, 6)	34	63	46	75	4	0	0
(1, 1, 1, 1, 1, 1, 2, 2, 3, 5)	110	138	133	173	16	0	0
(1, 1, 1, 1, 1, 1, 2, 2, 4, 4)	40	60	57	72	12	0	0
(1, 1, 1, 1, 1, 1, 2, 3, 3, 4)	76	102	101	75	3	0	0
(1, 1, 1, 1, 1, 1, 3, 3, 3, 3)	6	6	5	2	0	0	0
(1, 1, 1, 1, 1, 2, 2, 2, 2, 5)	20	58	55	36	1	0	0
(1, 1, 1, 1, 1, 2, 2, 2, 3, 4)	66	171	193	119	0	0	0
(1, 1, 1, 1, 1, 2, 2, 3, 3, 3)	21	74	90	39	0	0	0
(1, 1, 1, 1, 2, 2, 2, 2, 2, 4)	7	33	47	14	0	0	0
(1, 1, 1, 1, 2, 2, 2, 2, 3, 3)	15	68	94	12	0	0	0
(1, 1, 1, 2, 2, 2, 2, 2, 2, 3)	1	14	17	0	0	0	0
(1, 1, 2, 2, 2, 2, 2, 2, 2, 2)	1	2	1	0	0	0	0

Chapter 4
Uncertainty of Network Structure Identification

Abstract In this chapter, we develop a theoretical foundation to study uncertainty of network structure identification algorithms. We suggest to consider identification algorithms as multiple decision statistical procedures. In this framework, uncertainty is connected with a loss function and it is defined as expected value of the total loss, known as the risk function. We discuss, from this point of view, different measures of quality of multiple hypotheses testing and binary classification. We emphasize the class of additive loss functions as the most appropriate for network structure identification. We show that under some additional conditions, the risk function for additive losses is a linear combination of expected value of the numbers of Type I (False Positive) and Type II (False Negative) errors.

4.1 Multiple Decision Approach

In this section, we consider the network structure identification problem in the framework of decision theory [60, 91]. According to this approach, we specify the decision procedure (decision rule) and risk function. Let (X, γ) be a random variable network, (V, Γ) be the network model generated by (X, γ). Assume that vector X has distribution from some class K. Each distribution P_θ from the class K is associated with some parameter θ from the parameter space Ω. The network structure is an unweighted graph (U, E), where $U \subset V$, and E is a set of edges between nodes in U. For better understanding, we develop our approach for the networks structures with $U = V$ (threshold graph, concentration graph, maximum spanning tree, planar maximally filtered graph). The case $U \neq V$ can be treated in a similar way. For the case $U = V$, network structure is defined by an adjacency matrix $(N \times N)$.

Define the set $\mathcal{G}$ of all $N \times N$ symmetric matrices $G = (g_{i,j})$ with $g_{i,j} \in \{0, 1\}$, $i, j = 1, 2, \ldots, N$, $g_{i,i} = 0, i = 1, 2, \ldots, N$. Matrices $G \in \mathcal{G}$ represent adjacency matrices of all simple undirected graphs with N vertices. The total number of matrices in $\mathcal{G}$ is equal to $L = 2^M$ with $M = N(N-1)/2$. The network structure

© The Author(s) 2020
V. A. Kalyagin et al., *Statistical Analysis of Graph Structures in Random Variable Networks*, SpringerBriefs in Optimization,
https://doi.org/10.1007/978-3-030-60293-2_4

identification problem can be formulated as a multiple decision problem of the choice of one from L hypotheses:

$$H_G : \text{G is the adjacency matrix of the true network structure} \tag{4.1}$$

Let $x = (x_i(t)) \in R^{N \times n}$ be a sample of the size n from the distribution X. The multiple decision statistical procedure $\delta(x)$ is a map from the sample space $R^{N \times n}$ to the decision space $D = \{d_G, G \in \mathscr{G}\}$, where the decision d_G is the acceptance of hypothesis $H_G, G \in \mathscr{G}$.

According to [91], the quality of the statistical procedure is defined by the risk function. Let $w(S, Q)$ be the loss from decision d_Q when hypothesis H_S is true, i.e.,

$$w(H_S; d_Q) = w(S, Q), \quad S, Q \in \mathscr{G}.$$

We assume that $w(S, S) = 0$. The risk function is defined by

$$\text{Risk}(S, \theta; \delta) = \sum_{Q \in \mathscr{G}} w(S, Q) P_\theta(\delta(x) = d_Q), \quad \theta \in \Omega_S,$$

where Ω_S is the parametric region corresponding to the hypothesis H_S (i.e., the set of distributions such that the true network structure in (V, Γ) has adjacency matrix S), and $P_\theta(\delta(x) = d_Q)$ is the probability that decision d_Q is taken.

The multiple decision statistical procedure can be represented by the matrix

$$\Phi(x) = \begin{pmatrix} 0 & \varphi_{1,2}(x) & \cdots & \varphi_{1,N}(x) \\ \varphi_{2,1}(x) & 0 & \cdots & \varphi_{2,N}(x) \\ \cdots & \cdots & \cdots & \cdots \\ \varphi_{N,1}(x) & \varphi_{N,2}(x) & \cdots & 0 \end{pmatrix}, \tag{4.2}$$

where $\varphi_{i,j}(x) \in \{0, 1\}$. In this case one has

$$\delta(x) = d_G, \text{ iff } \Phi(x) = G \tag{4.3}$$

The value $w(S, \Phi(x))$ is the loss from the decision d_Φ, when the true decision is d_S. Risk is the expected value of this loss for the fixed S and $\theta \in \Omega_S$. Uncertainty of multiple decision procedure can therefore be measured by the risk. This framework allows to study general statistical properties of network structure identification algorithms. In this book, we are interested in two properties: optimality and robustness.

For a given loss function, decision procedure δ is called **optimal in the class of decision procedures** $\mathscr{F}$ if

$$\text{Risk}(S, \theta; \delta) \leq \text{Risk}(S, \theta; \delta'), \quad \theta \in \Omega_S, \ S \in \mathscr{G}, \ \delta, \delta' \in \mathscr{F} \tag{4.4}$$

To find an optimal procedure, one needs to solve a multiobjective optimization problem. It is possible that this problem does not have a solution in the class $\mathscr{F}$. It is interesting to specify the class of procedures where there is an optimal procedure. In this book, we consider the class of unbiased procedures and describe an optimal procedure in this class for some network structure identification problem. In the general case, one can look for Pareto optimal or admissible procedures.

For a given loss function, decision procedure is called **distribution free (robust) in the class of distributions K** if

$$\text{Risk}(S, \theta; \delta) = \text{Risk}(S, \theta'; \delta), \ P_\theta, P_{\theta'} \in K, \ \theta, \theta' \in \Omega_S \tag{4.5}$$

Such procedures are especially important for practical use in the case where there is no enough information about the distribution. In this book, we propose a new type of identification procedures which are robust (distribution free) in the class of elliptical distributions.

4.2 Loss and Risk Functions

The multiple decision approach allows to consider in the same framework many measures of uncertainty (error rate) widely used in multiple hypotheses testing and machine learning. In this section, we consider the most used among them. Let $S = (s_{i,j})$, $Q = (q_{i,j})$ be two adjacency matrix from $\mathscr{G}$. Loss function $w(S, Q)$ must be related with the difference between S and Q. Depending on how this difference is measured, one has different measures of uncertainty (error rates).

The most simple loss function is

$$w_{\text{Simple}}(S, Q) = \begin{cases} 1 \ if \ S \neq Q \\ 0 \ if \ S = Q \end{cases}$$

The associated risk is the probability of the false decision $\text{Risk}(S, \theta; \delta) = P_\theta(\delta(x) \neq d_S)$, $\theta \in \Omega_S$. This loss function does not take into account how large is the difference between the matrices, only the fact of the difference is indicated. Moreover, this loss function does not make difference between two types of error: zero is replaced by one (Type I error), or one is replaced by zero (Type II error). However, in practice it can be important. To explain better what follows we introduce two tables. Table 4.1 illustrates Type I and Type II errors for the individual edge (i, j).

Table 4.2 illustrates the difference between S and Q based on the numbers of Type I and Type II errors.

Table 4.1 represents all possible cases for different values of $s_{i,j}$ and $q_{i,j}$. Value 0 means that the edge (i, j) is not included in the corresponding structure, value 1 means that the edge (i, j) is included in the corresponding structure. We associate

Table 4.1 Type I and Type II errors for the edge (i, j)

$q_{i,j} \backslash s_{i,j}$	0	1
0	Edge is not included correctly, no error	Edge is not included incorrectly, Type II error
1	Edge is included incorrectly, Type I error	Edge is included correctly, no error

Table 4.2 Numbers of Type I and Type II errors

$Q \backslash S$	0 in S	1 in S	Total
0 in Q	TN	FN	Number of 0 in Q
1 in Q	FP	TP	Number of 1 in Q
Total	Number of 0 in S	Number of 1 in S	$N(N-1)/2$

the case $s_{i,j} = 0, q_{i,j} = 1$ with Type I error (false edge inclusion), and we associate the case $s_{i,j} = 1, q_{i,j} = 0$ with Type II error (false edge noninclusion). Table 4.2 represents the numbers of Type I errors (False Positive, FP), number of Type II errors (False Negative, FN), and numbers of correct decisions (True Positive, TP and True Negative, TN).

In multiple hypotheses testing, important attention is paid to the control of FWER (Family Wise Error Rate). FWER is the probability of at least one Type I error [9, 34, 62]. In the framework of multiple decision approach, it is related with the following loss and risk functions:

$$w_{\text{FWER}(S,Q)} = \begin{cases} 1 \; if \; FP > 0 \\ 0 \; if \; FP = 0 \end{cases}$$

In this case, one has

$$\text{Risk}_{\text{FWER}(S,\theta;\delta)} = E_\theta(w_{\text{FWER}(S;\delta)}) = P_\theta(FP(S;\delta) > 0) = \text{FWER}$$

These loss and associated risk functions take into account only one type of errors (Type I) and do not take into account the numbers of such errors. It was noted that Bonferroni and Holm procedures control FWER for any S in the following sense: FWER $\leq \alpha$. Natural generalization of FWER is k-FWER [61]. To obtain k-FWER, one can define the following loss function:

$$w_{k-\text{FWER}}(S, Q) = \begin{cases} 1 \; if \; FP \geq k \\ 0 \; if \; FP < k \end{cases}$$

For this loss function, one has

$$\text{Risk}_{k-\text{FWER}}(S, \theta; \delta) = E_\theta(w_{k-\text{FWER}}(S; \delta)) = P_\theta(FP(S; \delta) \geq k) = \text{k-FWER}$$

It is known that some modifications of Bonferroni and Holm procedures control k-FWER for any S [61]. In multiple hypotheses testing, Type II errors are often taken into account using a generalization of the classical notion of the power of test. We consider Conjunctive Power (CPOWER) and Disjunctive Power (DPOWER) [9]. Conjunctive power is analogous to FWER for the Type II errors. It is defined as probability of absence of Type II error (all edges in S are included in the Q). It can be obtained with the use of the following loss function:

$$w_{\text{CPOWER}(S,Q)} = \begin{cases} 1 \ if \ FN > 0 \\ 0 \ if \ FN = 0 \end{cases}$$

The associated risk function can be calculated as

$$\text{Risk}_{\text{CPOWER}(S,\theta;\delta)} = E_\theta(w_{\text{CPOWER}(S;\delta)}) = P_\theta(FN(S;\delta) > 0) =$$

$$= 1 - P_\theta(FN(S;\delta) = 0) = 1 - \text{CPOWER}$$

These loss and risk functions take into account only one type of errors (Type II) and do not take into account the numbers of such errors. Alternatively, Disjunctive Power is defined as the probability of at least one correct edge inclusion (at least one edge in S is included in Q). Define the loss function

$$w_{\text{DPOWER}(S,Q)} = \begin{cases} 1 \ if \ TP = 0 \\ 0 \ if \ TP > 0 \end{cases}$$

In this case, one has

$$\text{Risk}_{\text{DPOWER}(S,\theta;\delta)} = E_\theta(w_{\text{DPOWER}(S;\delta)}) = P_\theta(TP(S;\delta) = 0) =$$

$$= 1 - P_\theta(TP(S;\delta) > 0) = 1 - \text{DPOWER}$$

Considered measures of uncertainty, FWER, k-FWER, CPOWER, DPOWER do not take into account the numbers of associated errors. Next uncertainty measures take into account the numbers of errors. We start with Per-Family Error Rate (PFER) and Per-Comparison Error Rate (PCER). PFER is defined as the expected number of Type I errors (FP). Associated loss function can be defined as $w_{PFER} = FP$. PCER is defined by $PCER = PFER/M$, $M = N(N - 1)/2$. Average Power (AVE) is defined by $E(TP/(FN + TP))$. Associated loss function can be defined as $w_{\text{AVE}} = FN/(TP + FN)$. In this case, one has $\text{Risk}_{\text{AVE}} = 1 - \text{AVE}$. In binary classification, Risk_{AVE} is related with False Negative Rate, and $1 - \text{Risk}_{\text{AVE}}$ is related with Sensitivity or Recall. All these uncertainty characteristics take into account only one type of errors.

Both type and numbers of errors are taken into account in False Discovery Rate (FDR) and Accuracy. FDR is defined as $\text{FDR} = E(FP/(FP + TP))$. Associated loss function can be defined as $w_{FDR} = FP/(FP + TP)$. One has $\text{Risk}_{FDR} = \text{FDR}$. Accuracy (ACC), or proportion of correct decisions, is defined as

$ACC = E(TP+TN)/M$, $M = N(N-1)/2$. It can be defined by the following loss function $w_{ACC} = (FP+FN)/M$, $M = N(N-1)/2$. One has $\text{Risk}_{ACC} = 1-\text{ACC}$. There are many other uncertainty characteristics which are used in the literature. All of them are related with TN, FN, TP, FP. For example, ROC curve is defined in the coordinates False Positive Rate ($FP/(FP + TN)$) and True Positive Rate ($TP/(TP + FN)$), and Recall-Precision curve is defined in the coordinates Recall ($TP/(TP + FN)$), Precision ($TP/(TP + FP)$). FDR is one of the most popular measures of uncertainty both in multiple testing and machine learning. However, this measure has some drawbacks. For example, if the true structure is sparse (the number of zeros in S is large) then FDR can be close to 1 (bad classifier), but accuracy of both class classification can be close to 1. Similarly, in the case when the true structure is dense (the number of ones in S is large), FDR can be close to zero, but accuracy of both class classification can be close to 0.

4.3   Additive Loss and Risk Functions

In this book, we suggest to attract attention to additive loss functions. Define the individual loss for the edge (i, j) as

$$
w_{i,j}(S, Q) = \begin{cases} a_{i,j}, & \text{if } s_{i,j} = 0, q_{i,j} = 1, \\ b_{i,j}, & \text{if } s_{i,j} = 1, q_{i,j} = 0, \\ 0, & \text{otherwise} \end{cases}
$$

$a_{i,j}$ is the loss from the false inclusion of edge (i, j) in the structure Q, and $b_{i,j}$, is the loss from the false noninclusion of the edge (i, j) in the structure Q, $i, j = 1, 2, \ldots, N$; $i \neq j$. Following Lehmann [60], we call the loss function $w(S, Q)$ additive if

$$
w(S, Q) = \sum_{i=1}^{N} \sum_{j=1}^{N} w_{i,j}(S, Q) \tag{4.6}
$$

In this case, the total loss from the misclassification of S is equal to the sum of losses from the misclassification of individual edges:

$$
w(S, Q) = \sum_{\{i,j: s_{i,j}=0; q_{i,j}=1\}} a_{i,j} + \sum_{\{i,j: s_{i,j}=1; q_{i,j}=0\}} b_{i,j}
$$

This loss function takes into account both types of individual errors and highlights importance of each type of error. The following statement is true:

Theorem 4.1 *Let the loss function w be defined by (4.6), and $a_{i,j} = a$, $b_{i,j} = b$, $i \neq j$, $i, j = 1, 2, \ldots, N$. Then*

$$\text{Risk}(S, \theta; \delta) = a E_\theta[Y_I(S, \delta)] + b E_\theta[Y_{II}(S, \delta)], \quad \theta \in \Omega_S$$

where $Y_I(S, \delta) = FP(S; \delta)$, $Y_{II}(S, \delta) = FN(S; \delta)$ are the numbers of Type I and Type II errors by δ when the state (true decision) is d_S.

Proof One has

$$\text{Risk}(S, \theta; \delta) = \sum_{Q \in \mathscr{G}} w(S, Q) P_\theta(\delta(x) = d_Q) =$$

$$= \sum_{Q \in \mathscr{G}} [\sum_{\{i,j : s_{i,j}=0; q_{i,j}=1\}} a_{i,j} + \sum_{\{i,j : s_{i,j}=1; q_{i,j}=0\}} b_{i,j}] P_\theta(\delta(x) = d_Q).$$

$$= \sum_{Q \in \mathscr{G}} [a Y_I(S; \delta) + b Y_{II}(S; \delta)] P_\theta(\delta(x) = d_Q) = a E_\theta[Y_I(S, \delta)] + b E_\theta[Y_{II}(S, \delta)]$$

The theorem is proved.

Some important measures of uncertainty described above correspond to additive loss functions. In particular, for $a = b = 1/M$, $M = N(N-1)/2$, this loss function is related with ACC (Accuracy). Note that by the choice of a, b one can take into account the numbers of elements in unbalanced classes. For $a = 1$, $b_{i,j} = 0$, one get PFER as the associated risk function. Therefore, in our opinion, a risk function associated with additive losses can be well adapted to measure uncertainty of network structures identification algorithms.

Chapter 5
Robustness of Network Structure Identification

Abstract In this chapter, we discuss robustness of network structure identification algorithms. We understand robustness of identification algorithm as the stability of the risk function with respect to the distribution of the vector X from some class of distributions (distribution free property). We show that popular identification algorithms based on sample Pearson correlations are not robust in the class of elliptical distributions. To overcome this issue, we consider the sign similarity network, introduce a new class of network structure identification algorithms, and prove its robustness in the class of elliptical distributions. We show how to use these algorithms to construct robust network structure identification algorithms in other correlation networks.

5.1 Concept of Robustness

Let (X, γ) be a random variable network, (V, Γ) be the network model generated by (X, γ). Assume that vector X has distribution from some class K and for all $X \in K$ the network models (V, Γ) are identical. It means that a given network structure (U, E) is the same in all network models (V, Γ), $X \in K$. Let δ be an identification procedure (identification algorithm) for the network structure (U, E). Let $S \in \mathscr{G}$ be adjacency matrix corresponding to the network structure (U, E). Consider the loss function $w(S, Q)$. The associated risk function is defined as (see Chap. 4):

$$R(S, \theta; \delta) = \sum_{Q \in \mathscr{G}} w(S, Q) P_\theta(\delta = d_Q), \quad \theta \in \Omega_S$$

The true network structure S is the same for all $X \in K$, but the quality (risk function) of the identification procedure δ can depend on distribution of $X \in K$. The decision procedure δ is called *robust (stable or distribution free) in the class of distributions K* if

© The Author(s) 2020
V. A. Kalyagin et al., *Statistical Analysis of Graph Structures in Random Variable Networks*, SpringerBriefs in Optimization,
https://doi.org/10.1007/978-3-030-60293-2_5

$$\text{Risk}(S, \theta; \delta) = \text{Risk}(S, \theta'; \delta), \ \ P_\theta, P_{\theta'} \in K. \tag{5.1}$$

Uncertainty of distribution free procedure does not depend on distribution from the class K, network model (V, Γ) being fixed. Such procedures are especially important for practical use in the case where there is not enough information about the distribution. In this book, we consider the class of elliptical distributions, defined by the density function:

$$f(x; \mu, \Lambda) = |\Lambda|^{-\frac{1}{2}} g\{(x - \mu)' \Lambda^{-1} (x - \mu)\} \tag{5.2}$$

where $\Lambda = (\lambda_{i,j})_{i,j=1,2,\dots,N}$ is positive definite symmetric matrix, $g(x) \geq 0$, and

$$\int_{-\infty}^{\infty} \cdots \int_{-\infty}^{\infty} g(y'y) dy_1 dy_2 \cdots dy_N = 1$$

Let $K(\Lambda)$ be the class of elliptical distributions with fixed matrix Λ. It follows from the Chap. 2 that for $X \in K(\Lambda)$ the network models (V, Γ) are identical in each random variables network: Pearson correlation network, Fechner correlation network, Kruskal correlation network, Kendall correlation network, sign similarity network, and partial correlation network. Therefore, one can investigate the robustness of network structure identification procedures in all these networks for distributions $X \in K(\Lambda)$. In addition, network models in different random variables networks are related. It implies that an identification procedure in one network can be used as identification procedure for corresponding network structure in other network.

Let the measure of similarity γ be fixed. Consider a random variables networks with $X \in K(\Lambda)$. To investigate the robustness of network structure identification algorithms, we compare the risk functions for different distributions $X \in K(\Lambda)$. Variation of the risk function with variation of distribution is an indicator of nonrobustness of the identification procedure. Nonvariation of the risk function with variation of distribution is an indicator of robustness. To prove the robustness, one needs a theoretical arguments; to prove the nonrobustness, it is enough to present an example.

Example 5.1 This example shows that Kruskal algorithm based on sample Pearson correlations for MST identification is not robust, i.e., its risk function essentially depends on distribution from the class $K(\Lambda)$. Consider the Pearson correlation random variables network with $N = 10$, $X \in K(\Lambda)$, and matrix Λ is given below

Fig. 5.1 Maximum spanning
tree for Example 5.1

$$\begin{pmatrix} & 1 & 2 & 3 & 4 & 5 & 6 & 7 & 8 & 9 & 10 \\ 1 & 1.0000 & 0.4312 & 0.3470 & 0.3420 & 0.3798 & 0.3935 & 0.3212 & 0.4928 & 0.3342 & 0.1341 \\ 2 & 0.4312 & 1.0000 & 0.4548 & 0.3994 & 0.4889 & 0.5377 & 0.6053 & 0.4597 & 0.5743 & 0.1167 \\ 3 & 0.3470 & 0.4548 & 1.0000 & 0.3571 & 0.4930 & 0.5322 & 0.4478 & 0.4096 & 0.4618 & 0.2404 \\ 4 & 0.3421 & 0.3994 & 0.3571 & 1.0000 & 0.4315 & 0.4095 & 0.3741 & 0.4196 & 0.3705 & 0.0676 \\ 5 & 0.3798 & 0.4889 & 0.4930 & 0.4315 & 1.0000 & 0.5364 & 0.4732 & 0.4202 & 0.4361 & 0.1596 \\ 6 & 0.3935 & 0.5377 & 0.5322 & 0.4095 & 0.5364 & 1.0000 & 0.5197 & 0.5133 & 0.4958 & 0.2200 \\ 7 & 0.3212 & 0.6053 & 0.4478 & 0.3741 & 0.4732 & 0.5197 & 1.0000 & 0.4465 & 0.4866 & 0.1512 \\ 8 & 0.4929 & 0.4597 & 0.4096 & 0.4196 & 0.4202 & 0.5133 & 0.4465 & 1.0000 & 0.4686 & 0.1690 \\ 9 & 0.3342 & 0.5743 & 0.4618 & 0.3705 & 0.4361 & 0.4958 & 0.4866 & 0.4686 & 1.0000 & 0.1688 \\ 10 & 0.1341 & 0.1167 & 0.2404 & 0.0676 & 0.1596 & 0.2200 & 0.1512 & 0.1690 & 0.1688 & 1.0000 \end{pmatrix}$$

The associated network model (V, Γ) is given by $V = \{1, 2, \ldots, 10\}$, $\Gamma = \Lambda$.
True MST for this network model is presented in Fig. 5.1

Consider, as in example 3.4, the vector of degrees of vertices of MST ordered
in ascending order. True vector is $(1, 1, 1, 1, 1, 2, 2, 2, 3, 4)$. Let us introduce the
following loss function

$$w = \begin{cases} 1 \text{ if the vector of degrees of vertices of S differs from this vector of Q} \\ 0 \text{ if the vector of degrees of vertices of S is equal to this vector of Q} \end{cases}$$

The associated risk function is the probability of error in the identified vector of
degrees of vertices of MST ordered in ascending order. To study the variation of the
risk function with variation of distribution, we consider the mixture of two elliptical
distributions:

$$f_\epsilon(x) = (1 - \epsilon) f_{\text{Gauss}(x)} + \epsilon f_{\text{Student}(x)}, \quad 0 \le \epsilon \le 1$$

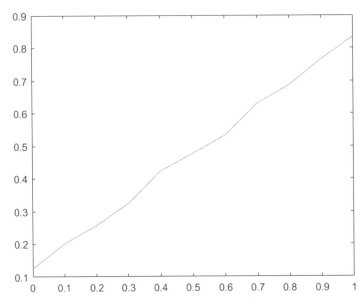

Fig. 5.2 Values of the risk function for different values of mixture parameter $\epsilon \in [0, 1]$ for Example 5.1

where f_{Gauss} is the density function for the multivariate Gaussian distribution from the class Gauss$(0, \Lambda)$, and f_{Student} is the density function for the multivariate Student distribution from the class Student$(0, \Lambda; \nu)$. We conduct the following experiment:

- Generate $n = 10\,000$ observations from the distribution with the density function $f_\epsilon(x)$ (degree of freedom for Student distribution $\nu = 3$).
- Calculate the estimations $\hat{\gamma}_{i,j}^P$ of Pearson correlations and construct the matrix $\hat{\Gamma}$.
- Use the Kruskal algorithm to calculate sample MST in the network model $(V, \hat{\Gamma})$
- Calculate the vector of degrees of vertices of sample MST ordered in ascending order.
- Compare the vector of degrees of vertices of true MST with the vector of degrees of vertices of sample MST and calculate the value of the loss function w.
- Repeat the experiment $S = 1000$ times, and estimate the value of the risk function.

Result are shown on the Fig. 5.2.

As one can see, the risk function essentially varies with the variation of ϵ from 0 to 1. This means that the Kruskal algorithm for MST identification in Pearson correlation network is not robust.

Remark Correlation matrix for Example 5.1 was calculated from the data of Indian stock market for the year 2011. The correlation matrix was considered as true correlation matrix.

The following example gives a motivation to theoretical study of the robustness of network structure identification algorithms in sign similarity network.

Example 5.2 Consider the sign similarity networks for the class of distributions $K(\Lambda)$, where the matrix Λ is taken from the Example 5.1. Consider the Kruskal algorithm for MST identification in sign similarity network. True MST is the same, as in the Example 5.1. Consider the same loss function

$$
w = \begin{cases} 1 \text{ if the vector of degrees of vertices of S differs from this vector of Q} \\ 0 \text{ if the vector of degrees of vertices of S is equal to this vector of Q} \end{cases}
$$

The associated risk function is the probability of error in the identified vector of degrees of vertices of MST ordered in ascending order. To study the variation of the risk function with variation of distribution, we consider the same mixture of two elliptical distributions:

$$
f_\epsilon(x) = (1 - \epsilon) f_{\text{Gauss}(x)} + \epsilon f_{\text{Student}(x)}, \quad 0 \le \epsilon \le 1
$$

where f_{Gauss} is the density function for the multivariate Gaussian distribution from the class Gauss$(0, \Lambda)$, and f_{Student} is the density function for the multivariate Student distribution from the class Student$(0, \Lambda; \nu)$. We conduct the similar experiment:

- Generate $n = 10\,000$ observations from the distribution with the density function $f_\epsilon(x)$ (degree of freedom for Student distribution $\nu = 3$).
- Calculate the estimations $\hat{\gamma}_{i,j}^{Sg}$ of sign similarities and construct the matrix $\hat{\Gamma}$.
- Use the Kruskal algorithm to calculate sample MST in the network model $(V, \hat{\Gamma})$
- Calculate the vector of degrees of vertices of sample MST ordered in ascending order.
- Compare the vector of degrees of vertices of true MST with the vector of degrees of vertices of sample MST and calculate the value of the loss function w.
- Repeat the experiment $S = 1000$ times, and estimate the value of the risk function.

Results are shown in Fig. 5.3.

As one can see, the risk function is almost stable with the variation of ϵ from 0 to 1.

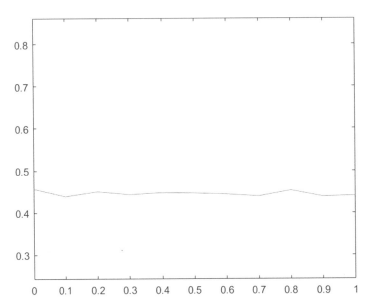

Fig. 5.3 Values of the risk function for different values of mixture parameter $\epsilon \in [0, 1]$ for Example 5.2

5.2   Robust Network Structure Identification in Sign Similarity Network

In this section, we prove that the network structure identification algorithms in sign similarity networks described in the Chap. 3 are robust (distribution free) in the class $K(\Lambda)$ of elliptical distributions.

The sign similarity network is a pair (X, γ^{Sg}), where γ^{Sg} is the probability of sign coincidence for a pair of random variables. Associated network model is (V, Γ), where $\Gamma = (\gamma_{i,j})$,

$$\gamma_{i,j} = \gamma^{Sg}(X_i, X_j) = P((X_i - E(X_i))(X_j - E(X_j)) > 0)$$

For $X \in K(\Lambda)$, all network models (V, Γ) coincide. Therefore, all true network structures coincide too. Network structure identification algorithms in sign similarity network are described in Chap. 3. They are based on the following statistics

$$T_{i,j}^{Sg} = \sum_{t=1}^{n} I_{i,j}^{Sg}(t) \tag{5.3}$$

where

$$I_{i,j}^{Sg}(t) = \begin{cases} 1, & (x_i(t) - \mu_i)(x_j(t) - \mu_j) \geq 0 \\ 0, & (x_i(t) - \mu_i)(x_j(t) - \mu_j) < 0 \end{cases}$$

We start with two lemmas.

Lemma 5.1 *Let random vector $(X_1, \ldots, X_N)$ has elliptical distribution with density*

$$f(x; 0, \Lambda) = |\Lambda|^{-1/2} g(x' \Lambda x)$$

Then the probabilities

$$p(i_1, i_2, \ldots, p_N) := P(i_1 X_1 > 0, i_2 X_2 > 0, \ldots, i_N X_N > 0) \tag{5.4}$$

do not depend on the function g for any $i_k \in \{-1, 1\}$, $k = 1, 2, \ldots, N$.

Proof One has

$$P(i_1 X_1 > 0, i_2 X_2, \ldots, i_N X_N > 0) = \int_{i_k x_k > 0, k=1,2,\ldots,N} |\Lambda|^{-\frac{1}{2}} g(x' \Lambda x) dx_1 \ldots dx_N \tag{5.5}$$

Matrix Λ is symmetric positive definite; therefore, there exists a matrix C such that $C' \Lambda C = I$. Put $y = C^{-1} x$. Then $x = Cy$ and

$$\int_{i_k x_k > 0, k=1,2,\ldots,N} |\Lambda|^{-\frac{1}{2}} g(x' \Lambda x) dx_1 \ldots dx_N = \int_D g(y' y) dy_1 \ldots dy_N \tag{5.6}$$

where D is given by

$$0 < i_k (c_{k,1} y_1 + c_{k,2} y_2 + \ldots + c_{k,N} y_N) < \infty, \quad k = 1, 2, \ldots, N \tag{5.7}$$

The vector y can be written in polar coordinates as:

$$\begin{aligned}
y_1 &= r \sin(\theta_1) \\
y_2 &= r \cos(\theta_1) \sin(\theta_2) \\
y_3 &= r \cos(\theta_1) \cos(\theta_2) \sin(\theta_3) \\
&\ldots \\
y_{N-1} &= r \cos(\theta_1) \cos(\theta_2) \ldots \cos(\theta_{N-2}) \sin(\theta_{N-1}) \\
y_N &= r \cos(\theta_1) \cos(\theta_2) \ldots \cos(\theta_{N-2}) \cos(\theta_{N-1})
\end{aligned} \tag{5.8}$$

where $-\frac{\pi}{2} \leq \theta_i \leq \frac{\pi}{2}, i = 1 \ldots, N - 2; -\pi \leq \theta_{N-1} \leq \pi, 0 \leq r \leq \infty$
The Jacobian of the transformation (5.8) is

$$r^{N-1} \cos^{N-2}(\theta_1) \cos^{N-3}(\theta_2) \ldots \cos(\theta_{N-2})$$

In polar coordinates region, (5.7) is transformed to the region $D' \times R_+^1$ where D' is given by ($k = 1, 2, \ldots, N$):

$$0 < i_k(c_{11}\sin(\theta_1) + \ldots + c_{1N}\cos(\theta_1)\cos(\theta_2)\ldots\cos(\theta_{N-2})\cos(\theta_{N-1})) < \infty$$
$$(5.9)$$

Then $p(i_1, i_2, \ldots, i_N)$ can be written as

$$\int_{D'}\int_0^\infty r^{N-1}\cos^{N-2}(\theta_1)\cos^{N-3}(\theta_2)\ldots\cos(\theta_{N-2})g(r^2)drd\theta_1\ldots d\theta_{N-1} =$$

$$= \int_{D'}\cos^{N-2}(\theta_1)\cos^{N-3}(\theta_2)\ldots\cos(\theta_{N-2})d\theta_1\ldots d\theta_{N-1}\int_0^\infty r^{N-1}g(r^2)dr$$

It is known [1] that

$$\int_0^\infty r^{N-1}g(r^2)dr = \frac{1}{C(N)}$$

where

$$C(N) = \int_{-\frac{\pi}{2}}^{\frac{\pi}{2}}\ldots\int_{-\frac{\pi}{2}}^{\frac{\pi}{2}}\int_{-\pi}^{\pi}\cos^{N-2}(\theta_1)\cos^{N-3}(\theta_2)\ldots\cos(\theta_{N-2})d\theta_1\ldots d\theta_{N-1}$$

The region D' is defined by the matrix Λ and does not depend on the function g. Therefore, the probabilities $p(i_1, i_2, \ldots, i_N)$ are defined by the matrix Λ and do not depend on the function g. In particular for $N = 2$ the probabilities $P(X_1 > 0, X_2 > 0)$, $P(X_1 > 0, X_2 < 0)$, $P(X_1 < 0, X_2 > 0)$, $P(X_1 < 0, X_2 < 0)$ do not depend on the function g. The lemma is proved.

Next lemma shows that joint distribution of statistics $T_{i,j}^{Sg}$ does not depend on the function g too.

Lemma 5.2 *Let random vector $(X_1, \ldots, X_N)$ has elliptical distribution with density*

$$f(x; 0, \Lambda) = |\Lambda|^{-1/2}g(x'\Lambda x)$$

Then joint distribution of the statistics $T_{i,j}^{Sg}$ $(i, j = 1, 2, \ldots, N; i \neq j)$ does not depend on the function g.

Proof Statistic $T_{i,j}^{Sg}$ can be written as

$$T_{i,j}^{Sg} = \frac{1}{2} + \frac{1}{2}\sum_{t=1}^n \text{sign}(X_i(t))\text{sign}(X_j(t))$$

It follows from the Lemma 5.1 that joint distribution of the random vector $\text{sign}(X) = (\text{sign}(X_1), \text{sign}(X_2), \ldots, \text{sign}(X_N))$ is defined by the matrix Λ and does not depend on the function g. Random vectors $\text{sign}(X(t))$, $t = 1, 2, \ldots, n$ are independent and identically distributed. Therefore, the joint distribution of random variables $\text{sign}(X_i(t))$, $i = 1, 2, \ldots, N$, $t = 1, 2, \ldots, n$ is defined by the matrix Λ and does not depend on the function g. It implies that joint distribution of statistics $T_{i,j}^{Sg}$, $i, j = 1, 2, \ldots, N$; $i < j$ does not depend on the function g. The lemma is proved.

It is possible to give an explicit expression for the joint distribution of statistics $T_{i,j}^{Sg}$, $i, j = 1, 2, \ldots, N$; $i < j$ using probabilities (5.4). To simplify the notations, let us introduce statistics $T_1, T_2, \ldots, T_M$, $M = N(N-1)/2$ by

$$T_l = T_{k,s}^{Sg}, \quad l = l(k, s) = N(k-1) - \frac{1}{2}k(k-1) + s - k, \quad k < s, \quad l = 1, 2, \ldots, M$$

Then

$$P(T_1 = k_1, \ldots, T_M = k_M) = \sum_A n! \prod_{j_s \in \{0,1\}} \frac{q(j_1, j_2, \ldots, j_M)^{m(j_1, j_2, \ldots, j_M)}}{m(j_1, j_2, \ldots, j_M)!}$$

where

$$A = \{m(j_1, j_2, \ldots, j_M), j_s \in \{0, 1\} :$$

$$\sum_{j_s \in \{0,1\}} m(j_1, j_2, \ldots, j_M) = n; \sum_{j_s=1} m(j_1, j_2, \ldots, j_M) = k_s, s = 1, 2, \ldots, M\}$$

and

$$q(j_1, j_2, \ldots, j_M) = \sum_B p(i_1, i_2, \ldots, i_N)$$

where

$$B = \{(i_1, i_2, \ldots, i_N), i_s \in \{-1, 1\} : i_k = i_s, \text{ if } j_l = 1; \ i_k = -i_s, \text{ if } j_l = 0; l = l(k, s)\}$$

In particular, for $N = 2$ one has

$$P(T_1 = k_1) = \frac{n!}{k_1!(n-k_1)!} q(1)^{k_1} q(0)^{n-k_1}$$

where

$$q(1) = p(-1, -1) + p(1, 1), \quad q(0) = p(1, -1) + p(-1, 1), \quad q(1) + q(0) = 1$$

First we prove the robustness in the class $K(\Lambda)$ of threshold graph identification algorithms in sign similarity networks.

Theorem 5.1 *Let Λ be positive definite symmetric matrix, μ be a fixed vector. Then for any loss function, simultaneous inference, Holm, Hochberg, and Benjamini-Hochberg threshold graph identification procedures in sign similarity network are robust (distribution free) in the class of elliptical distributions $K(\Lambda)$.*

Proof Let γ_0^{Sg} be a given threshold. p-values of individual tests for threshold graph identification are defined by (see Sect. 3.1.5):

$$p^{Sg}(i, j) = 1 - F(T_{i,j}^{Sg}; n, \gamma_0^{Sg}) \tag{5.10}$$

where $F(t; n, \gamma_0^{Sg})$ is the cumulative distribution function for binomial distribution with parameters (n, γ_0^{Sg}). It follows from the Lemma 5.2 that the joint distribution of $p^{Sg}(i, j)$ does not depend on the function g. All procedures under consideration are based on two operations: ordering of $p^{Sg}(i, j)$ (which is the same as ordering of $T_{i,j}^{Sg}$), and comparison of $p^{Sg}(i, j)$ with the fixed constants (which is the same as comparison of $T_{i,j}^{Sg}$ with another fixed constants). For all procedures, the probabilities $P(\delta(x) = d_Q/H_S)$ are defined by the joint distribution of $p^{Sg}(i, j)$ (or by the joint distribution of $T_{i,j}^{Sg}$). It implies that these probabilities do not depend on g. More precisely, denote by δ^S, δ^H, δ^{Hg}, δ^{BH}, respectively, simultaneous inference, Holm, Hochberg, and Benjamini-Hochberg γ_0^{Sg} – threshold graph identification procedures.

Simultaneous inference One has for δ^S procedure:

$$P(\delta^S(x) = d_Q/H_S) = P_{\Lambda,g}(\Phi(x) = Q) =$$

$$= P_{\Lambda,g}(T_{i,j}^{Sg} > c_{i,j}, \text{ for } q_{i,j} = 1 \text{ and } T_{i,j}^{Sg} \le c_{i,j}, \text{ for } q_{i,j} = 0) =$$

$$= P_{\Lambda}(T_{i,j}^{Sg} > c_{i,j}, \text{ for } q_{i,j} = 1 \text{ and } T_{i,j}^{Sg} \le c_{i,j}, \text{ for } q_{i,j} = 0)$$

where $\Lambda \in H_S$. The theorem for the simultaneous inference procedure follows.

Holm procedure Let $I = \{(i, j) : q_{i,j} = 1, i, j = 1, 2, \ldots, N, i < j\}, k = |I|$. For the Holm procedure one has

$$P(\delta^H(x) = d_Q/H_S) =$$

$$\sum_{\sigma} P_{\Lambda,g}(A_\sigma \cap B_\sigma \cap C_I) = \sum_{\sigma} P_{\Lambda}(A_\sigma \cap B_\sigma \cap C_I)$$

where σ is the set of all permutations of the set I and

$$A_\sigma = \{x \in R^{N \times n} : T_{\sigma(1)}^{Sg}(x) \geq T_{\sigma(2)}^{Sg}(x) \geq \cdots \geq T_{\sigma(k)}^{Sg}(x)\}$$

$$B_\sigma = \{x \in R^{N \times n} : T_{\sigma(1)}^{Sg}(x) > c_1^H, T_{\sigma(2)}^{Sg}(x) > c_2^H, \cdots, T_{\sigma(k)}^{Sg}(x) > c_k^H\}$$

$$C_I = \{x \in R^{N \times n} : \max_{(i,j) \notin I} T_{i,j}^{Sg} \leq c_{k+1}^H\}$$

Hochberg procedure Let $J = \{(i, j) : q_{i,j} = 0, i, j = 1, 2, \ldots, N, i < j\}$, $m = |J|$. For the Hochberg procedure one has

$$P(\delta^{Hg}(x) = d_Q/H_S) =$$

$$\sum_\sigma P_{\Lambda,g}(D_\sigma \cap E_\sigma \cap F_J) = \sum_\sigma P_\Lambda(D_\sigma \cap E_\sigma \cap F_J)$$

where σ is the set of all permutations of the set J and

$$D_\sigma = \{x \in R^{N \times n} : T_{\sigma(1)}^{Sg}(x) \leq T_{\sigma(2)}^{Sg}(x) \leq \cdots \leq T_{\sigma(m)}^{Sg}(x)\}$$

$$E_\sigma = \{x \in R^{N \times n} : T_{\sigma(1)}^{Sg}(x) < c_1^{Hg}, T_{\sigma(2)}^{Sg}(x) < c_2^{Hg}, \cdots, T_{\sigma(m)}^{Sg}(x) < c_m^{Hg}\}$$

$$F_J = \{x \in R^{N \times n} : \min_{(i,j) \notin J} T_{i,j}^{Sg} \geq c_{m+1}^{Hg}\}$$

Benjamini-Hochberg procedure Let $J = \{(i, j) : q_{i,j} = 0, i, j = 1, 2, \ldots, N, i < j\}$, $m = |J|$. For the Benjamini-Hochberg procedure one has

$$P(\delta^{BH}(x) = d_Q/H_S) =$$

$$\sum_\sigma P_{\Lambda,g}(D_\sigma \cap E_\sigma \cap F_J) = \sum_\sigma P_\Lambda(D_\sigma \cap E_\sigma \cap F_J)$$

where σ is the set of all permutations of the set J and

$$D_\sigma = \{x \in R^{N \times n} : T_{\sigma(1)}^{Sg}(x) \leq T_{\sigma(2)}^{Sg}(x) \leq \cdots \leq T_{\sigma(m)}^{Sg}(x)\}$$

$$E_\sigma = \{x \in R^{N \times n} : T_{\sigma(1)}^{Sg}(x) < c_1^{BH}, T_{\sigma(2)}^{Sg}(x) < c_2^{BH}, \cdots, T_{\sigma(m)}^{Sg}(x) < c_m^{BH}\}$$

$$F_J = \{x \in R^{N \times n} : \min_{(i,j) \notin J} T_{i,j}^{Sg} \geq c_{m+1}^{BH}\}$$

The theorem is proved.

Second we prove the robustness in the class $K(\Lambda)$ of concentration graph identification algorithms in sign similarity networks.

Theorem 5.2 *Let Λ be positive definite symmetric matrix, μ be a fixed vector. Then for any loss function, simultaneous inference, Holm, Hochberg, and Benjamini-Hochberg concentration graph identification procedures in sign similarity network are robust (distribution free) in the class of elliptical distributions $K(\Lambda)$.*

Proof The concentration graph in sign similarity network is defined as follows: the edge (i, j) is included in the concentration graph if and only if $\gamma_{i,j}^{Sg} \neq \frac{1}{2}$. The p-values of individual tests for concentration graph identification are defined by (see Sect. 3.2):

$$p^{Sg}(i, j) = \begin{cases} 2F(T_{i,j}^{Sg}; n, \frac{1}{2}), & \text{if } T_{i,j}^{Sg} \leq \frac{n}{2} \\ 2(1 - F(T_{i,j}^{Sg}; n, \frac{1}{2})), & \text{if } T_{i,j}^{Sg} > \frac{n}{2} \end{cases} \tag{5.11}$$

where $F(t; n, \frac{1}{2})$ is the cumulative distribution function for binomial distribution with parameters $(n, \frac{1}{2})$. It follows from the Lemma 5.2 that the joint distribution of $p^{Sg}(i, j)$ does not depend on the function g. All procedures under consideration are based, as above, on two operations: ordering of $p^{Sg}(i, j)$, and comparison of $p^{Sg}(i, j)$ with the fixed constants. For all procedures, the probabilities $P(\delta(x) = d_Q/H_S)$ are defined by the joint distribution of $p^{Sg}(i, j)$. It implies that these probabilities don't depend on g. The theorem follows.

Finally, we prove the robustness in the class $K(\Lambda)$ of Kruskal MST identification algorithm in sign similarity networks.

Theorem 5.3 *Let Λ be positive definite symmetric matrix, μ be a fixed vector. Then for any loss function, Kruskal MST identification algorithm in sign similarity network is robust (distribution free) in the class of elliptical distributions $K(\Lambda)$.*

Proof The first step of the Kruskal MST identification algorithm is to order $\hat{\gamma}_{i,j}^{Sg}$ in descending order. Each ordering defines a decision d_Q, where Q is the adjacency matrix of the associated MST. Probability of any ordering is defined by the joint distribution of the statistics $T_{i,j}^{Sg}$. Therefore, the probabilities $P(Kruskal(x) = d_Q/H_S)$ do not depend on the function g. It implies that the risk function is stable with variation of distributions in the class $K(\Lambda)$, and the theorem follows.

5.3   Robust Network Structure Identification in Correlation Networks

Robust algorithms of network structure identification in a sign similarity network can be adapted to construct robust identification algorithms in other correlation networks. In this section, we construct robust identification algorithms in Pearson, Kruskal, Fechner, and Kendall correlation networks.

It was shown in the Sect. 2.3 that for $X \in K(\Lambda)$, the network models and network structures for Pearson, Kruskal, Fechner, Kendall, and sign similarity networks are related. In particular, one has

$$\gamma_{i,j}^{Fh} = \gamma_{i,j}^{Kr} = \gamma_{i,j}^{Kd} = 2\gamma_{i,j}^{Sg} - 1, \quad \gamma_{i,j}^{P} = \sin[\pi(\gamma_{i,j}^{Sg} - 0,5)] = -\cos(\pi \gamma_{i,j}^{Sg})$$

$$(5.12)$$

Therefore, for $X \in K(\Lambda)$, true concentration graph and true MST in all correlation networks are the same as true concentration graph and true MST in sign similarity network. It implies that any robust identification algorithm in sign similarity network generates a robust identification algorithm in any correlation network.

The true threshold graph in any correlation network (Pearson, Kruskal, Fechner, and Kendall correlation networks) is a true threshold graph in sign similarity network with an appropriate choice of the threshold. More precisely, the appropriate choices of the thresholds in sign similarity network in connection with given thresholds in other correlations networks are given by

$$\gamma_0^{Sg} = \frac{1}{2}(1 + \gamma_0^{Fh}) = \frac{1}{2}(1 + \gamma_0^{Kr}) = \frac{1}{2}(1 + \gamma_0^{Kd}), \quad \gamma_0^{Sg} = \frac{1}{2} + \frac{1}{\pi} \arcsin(\gamma_0^{P})$$

It implies that any robust identification algorithm for the threshold graph identification in sign similarity network generates a robust identification algorithm in any correlation network.

Chapter 6
Optimality of Network Structure Identification

Abstract In this chapter, we discuss optimality of network structure identification algorithms. We introduce a concept of optimality in the sense of minimization of the risk function. We investigate optimality of identification procedures for two problems: concentration graph identification and threshold graph identification. We restrict our study to the risk functions associated with additive losses. We prove optimality of simultaneous inference for concentration graph identification in Gaussian partial correlation network in the class of unbiased procedures. For threshold graph identification, we prove optimality of simultaneous inference in Gaussian Pearson correlation network in the class of statistical procedures invariant under scale/shift transformations. Finally, we prove optimality of simultaneous inference for threshold graph identification in sign similarity network.

6.1 Concept of Optimality

In this section, we discuss the concept of optimality related to multiple decision statistical procedures. Let Ω be the set of parameters. By Ω_S we denote the parametric region corresponding to hypothesis H_S. For all $\theta \in \Omega_S$, the associated network structures have the same adjacency matrix S. Let $S = (s_{i,j})$, $Q = (q_{i,j})$, $S, Q \in \mathcal{G}$. By $w(S, Q)$ we denote the loss from decision d_Q when hypothesis H_S is true, i.e.,

$$w(H_S; d_Q) = w(S, Q), \quad S, Q \in \mathcal{G}.$$

We assume that $w(S, S) = 0$. Let $S \in \mathcal{G}$. The risk function of $\delta(x)$ is defined by

$$\text{Risk}(S, \theta; \delta) = \sum_{Q \in \mathcal{G}} w(S, Q) P_\theta(\delta(x) = d_Q), \quad \theta \in \Omega_S,$$

where $P_\theta(\delta(x) = d_Q)$ is the probability that decision d_Q is taken.

© The Author(s) 2020
V. A. Kalyagin et al., *Statistical Analysis of Graph Structures in Random Variable Networks*, SpringerBriefs in Optimization,
https://doi.org/10.1007/978-3-030-60293-2_6

For a given loss function w, the decision procedure δ is called **optimal in the class of decision procedures** $\mathscr{F}$ (see Sect. 4.1) if

$$\text{Risk}(S, \theta; \delta) \leq \text{Risk}(S, \theta; \delta'), \quad \theta \in \Omega_S, \quad S \in \mathscr{G}, \quad \delta, \delta' \in \mathscr{F} \tag{6.1}$$

To find an optimal procedure, one needs to solve a multiobjective optimization problem. It is possible that this problem has no solution in the class $\mathscr{F}$. Therefore, it is important to specify the class of procedures where an optimal procedure exists. In this book, we consider the classes of unbiased and shift-scale invariant procedures and describe an optimal procedure in these classes for some network structure identification problem. In the general case one can be interested in Pareto optimal or admissible procedures.

6.2 W-Unbiasedness for Additive Loss Function

The statistical procedure $\delta(x)$ is referred to as w-unbiased [62] if

$$\text{Risk}(S, \theta; \delta) = E_\theta w(S; \delta) \leq E_\theta w(S'; \delta) = \text{Risk}(S', \theta; \delta), \tag{6.2}$$

for any $S, S' \in \mathscr{G}, \theta \in \Omega_S$. The unbiasedness of statistical procedure δ for a general risk function means that δ comes closer in expectation to the true decision than to any other decision. For an additive loss function (see Sect. 4.3) and for the case where $a_{i,j} = a, b_{i,j} = b$, the w-unbiasedness of the identification procedure δ means that

$$aE_\theta[Y_I(S, \delta)] + bE_\theta[Y_{II}(S, \delta)] \leq aE_\theta[Y_I(S', \delta)] + bE_\theta[Y_{II}(S', \delta)]$$

for any $S, S' \in \mathscr{G}, \theta \in \Omega_S$.

Unbiasedness of multiple testing procedure for an additive loss function implies unbiasedness of individual tests. To show this one can take S, S' such that S' and S differ only in two positions (i, j) and (j, i). In this case, one has

$$\text{Risk}(S, \theta; \delta) = 2\text{Risk}(s_{i,j}, \theta; \varphi_{i,j}) + \sum_{(k,l) \neq (i,j); (k,l) \neq (j,i)} \text{Risk}(s_{k,l}, \theta; \varphi_{k,l})$$

and:

$$\text{Risk}(S', \theta; \delta) = 2\text{Risk}(s'_{i,j}, \theta; \varphi_{i,j}) + \sum_{(k,l) \neq (i,j); (k,l) \neq (j,i)} \text{Risk}(s_{k,l}, \theta; \varphi_{k,l})$$

Therefore,

$$\mathrm{Risk}(s_{i,j}, \theta; \varphi_{i,j}) \leq \mathrm{Risk}(s'_{i,j}, \theta; \varphi_{i,j}).$$

One has

$$\mathrm{Risk}(s_{i,j}, \theta, \varphi_{i,j}) = \begin{cases} a_{i,j} P_\theta(\varphi_{i,j} = 1), & \theta \in \omega_{i,j} \\ b_{i,j} P_\theta(\varphi_{i,j} = 0), & \theta \in \omega_{i,j}^{-1} \end{cases}$$

and

$$\mathrm{Risk}(s'_{i,j}, \theta, \varphi_{i,j}) = \begin{cases} b_{i,j} P_\theta(\varphi_{i,j} = 0), & \theta \in \omega_{i,j} \\ a_{i,j} P_\theta(\varphi_{i,j} = 1), & \theta \in \omega_{i,j}^{-1} \end{cases}$$

It implies that

$$a_{i,j} P_\theta(\varphi_{i,j} = 1) \leq b_{i,j} P_\theta(\varphi_{i,j} = 0), \text{ if } \theta \in \omega_{i,j}$$

and:

$$b_{i,j} P_\theta(\varphi_{i,j} = 0) \leq a_{i,j} P_\theta(\varphi_{i,j} = 1), \text{ if } \theta \in \omega_{i,j}^{-1}$$

It implies usual unbiasedness of individual test $\varphi_{i,j}$ [62]:

$$E_\theta(\varphi_{i,j}) \leq \alpha_{i,j}, \ \forall \theta \in \omega_{i,j}; \quad E_\theta(\varphi_{i,j}) \geq \alpha_{i,j}, \ \forall \theta \in \omega_{i,j}^{-1}$$

where

$$\alpha_{i,j} = \frac{b_{i,j}}{a_{i,j} + b_{i,j}}$$

For the case $a_{i,j} = a$, $b_{i,j} = b$, by an appropriate choice of the matrix S' one can obtain the following bounds for expected numbers of Type I and Type II errors for any unbiased procedure

$$EFP = E(Y_I) \leq \frac{b}{a+b}(TN + FP), \quad EFN = E(Y_{II}) \leq \frac{b}{a+b}(TP + FN)$$

where $(TN + FP)$ is the number of off diagonal zeros in the matrix S, $TP + FN$ is the number of ones in S.

6.3 Optimal Concentration Graph Identification in Partial Correlation Network

6.3.1 Gaussian Graphical Model Selection

A Gaussian graphical model graphically represents the dependence structure of a Gaussian random vector. For the undirected graphical model, this dependence structure is associated with simple undirected graph. This graph is given by its adjacency matrix, which is symmetric matrix with $\{0, 1\}$ entries, where zero means conditional independence and one means conditional dependence of random variables. It has been found to be an effective method in different applied fields such as bioinformatics, error-control codes, speech and language recognition, and information retrieval [41, 90]. One of the main questions in graphical models is how to recover the structure of a Gaussian graph from observations and what are statistical properties of associated algorithms. This problem is called the Gaussian graphical model selection problem (GGMS). A comprehensive survey of different approaches to this problem is given in [17, 20].

One of the first approaches to GGMS for undirected graphs was based on covariance selection procedures [16, 21]. GGMS problem is still popular our days [49, 64, 68, 75, 77, 78]. Several selection algorithms are used for large dimensional graphical models. Some approaches to GGMS are related with multiple hypotheses testing [20]. Measures of quality in multiple testing include the FWER (family wise error rate), k-FWER, FDR (false discovery rate), and FDP (false discovery proportion) [62]. Procedures with control of such type errors for GGMS are investigated in [20]. However, general statistical properties such as unbiasedness and optimality of these procedures are not well known. In this section, we combine the tests of a Neyman structure for individual hypotheses with simultaneous inference and prove that the obtained multiple decision procedure is optimal in the class of unbiased procedures.

6.3.2 GGMS Problem Statement

Let $X = (X_1, X_2, \ldots, X_N)$ be a random vector with the multivariate Gaussian distribution from $N(\mu, \Sigma)$, where $\mu = (\mu_1, \mu_2, \ldots, \mu_N)$ is the vector of means and $\Sigma = (\sigma_{i,j})$ is the covariance matrix, $\sigma_{i,j} = \text{cov}(X_i, X_j)$, $i, j = 1, 2, \ldots, N$. Let $x(t)$, $t = 1, 2, \ldots, n$ be a sample of size n from the distribution of X. We assume that $n > N$, and that the matrix Σ is nonsingular. The case $n < N$ has a practical interest too [64], but it is not considered. The undirected Gaussian graphical model is an undirected graph with N nodes. The nodes of the graph are associated with the random variables $X_1, X_2, \ldots, X_N$, edge (i, j) is included in the graph if the random variables X_i, X_j are conditionally dependent [1, 59].

The Gaussian graphical model selection problem consists of the identification of a graphical model from observations.

The partial correlation $\rho^{i,j}$ of X_i, X_j given X_k, $k \in N(i, j) = \{1, 2, \ldots, N\} \setminus \{i, j\}$ is defined as the correlation of X_i, X_j in the conditional distribution of X_i, X_j given X_k, $k \in N(i, j)$. This conditional distribution is Gaussian with the correlation $\rho^{i,j}$. It implies that the conditional independence of X_i, X_j given X_k, $k \in N(i, j) = \{1, 2, \ldots, N\} \setminus \{i, j\}$ is equivalent to the equation $\rho^{i,j} = 0$. Therefore, the Gaussian graphical model selection is equivalent to simultaneous inference on hypotheses of pairwise conditional independence $\rho^{i,j} = 0$, $i \neq j$, $i, j = 1, 2, \ldots, N$. The problem of pairwise conditional independence testing has the form:

$$h_{i,j} : \rho^{i,j} = 0 \text{ vs } k_{i,j} : \rho^{i,j} \neq 0, \quad i \neq j, i, j = 1, 2, \ldots, N \tag{6.3}$$

According to [59], the partial correlation can be written as

$$\rho^{i,j} = -\frac{\sigma^{i,j}}{\sqrt{\sigma^{i,i} \sigma^{j,j}}},$$

where $\sigma^{i,j}$ are the elements of the inverse matrix $\Sigma^{-1} = (\sigma^{i,j})$, known as the concentration or precision matrix of X. Hence, (6.3) is equivalent to

$$h_{i,j} : \sigma^{i,j} = 0, \text{ vs } k_{i,j} : \sigma^{i,j} \neq 0, i \neq j, i, j = 1, 2, \ldots, N \tag{6.4}$$

6.3.3 Uniformly Most Powerful Unbiased Tests of Neyman Structure

To construct UMPU test for the problem (6.4), we use a test of Neyman structure for natural parameters of exponential family. Let $f(x; \theta)$ be the density of the exponential family:

$$f(x; \theta) = c(\theta) exp \left(\sum_{j=1}^{M} \theta_j T_j(x) \right) m(x) \tag{6.5}$$

where $c(\theta)$ is a function defined in the parameters space, $m(x)$, $T_j(x)$ are functions defined in the sample space, and $T_j(X)$ are the sufficient statistics for θ_j, $j = 1, \ldots, M$.

Suppose that hypothesis has the form:

$$h_j : \theta_j = \theta_j^0 \text{ vs } k_j : \theta_j \neq \theta_j^0, \tag{6.6}$$

where θ_j^0 is fixed.

The UMPU test for hypotheses (6.6) is (see [62], Ch. 4, theorem 4.4.1):

$$\varphi_j = \begin{cases} 0, & \text{if } c'_j(t_1, \ldots, t_{j-1}, t_{j+1}, \ldots, t_M) < t_j < c''_j(t_1, \ldots, t_{j-1}, t_{j+1}, \ldots, t_M) \\ 1, & \text{otherwise} \end{cases}$$

(6.7)

where $t_i = T_i(x), i = 1, \ldots, M$. The constants c'_j, c''_j are defined from the equations

$$\int_{c'_j}^{c''_j} f(t_j; \theta_j^0 | T_i = t_i, i = 1, \ldots, M; i \neq j) dt_j = 1 - \alpha$$

(6.8)

and

$$\int_{-\infty}^{c'_j} t_j f(t_j; \theta_j^0 | T_i = t_i, i = 1, \ldots, M; i \neq j) dt_j +$$
$$+ \int_{c''_j}^{+\infty} t_j f(t_j; \theta_j^0 | T_i = t_i, i = 1, \ldots, M; i \neq j) dt_j =$$
$$= \alpha \int_{-\infty}^{+\infty} t_j f(t_j; \theta_j^0 | T_i = t_i, i = 1, \ldots, M; i \neq j) dt_j$$

(6.9)

where $f(t_j; \theta_j^0 | T_i = t_i, i = 1, \ldots, M; i \neq j)$ is the density of conditional distribution of statistic T_j given $T_i = t_i, i = 1, 2, \ldots, N, i \neq j$, and α is the significance level of the test.

6.3.4 Uniformly Most Powerful Unbiased Test for Conditional Independence

Now we construct the UMPU test for testing hypothesis of conditional independence (6.4). Consider statistics

$$S_{k,l} = \frac{1}{n} \sum_{t=1}^{n} (X_k(t) - \overline{X_k})(X_l(t) - \overline{X_l}),$$

Joint distribution of statistics $S_{k,l}, k, l = 1, 2, \ldots, N, n > N$ is given by Wishart density function [1]:

$$f(\{s_{k,l}\}) = \frac{[\det(\sigma^{k,l})]^{n/2} \times [\det(s_{k,l})]^{(n-N-2)/2} \times \exp[-(1/2) \sum_k \sum_l s_{k,l}\sigma^{k,l}]}{2^{(Nn/2)} \times \pi^{N(N-1)/4} \times \Gamma(n/2)\Gamma((n-1)/2) \cdots \Gamma((n-N+1)/2)}$$

if the matrix $S = (s_{k,l})$ is positive definite, and $f(\{s_{k,l}\}) = 0$ otherwise. It implies that statistics $S_{k,l}$ are sufficient statistics for natural parameters $\sigma^{k,l}$. Wishart density function can be written as:

$$f(\{s_{k,l}\}) = C(\{\sigma^{k,l}\}) \exp[-\sigma^{i,j}s_{i,j} - \frac{1}{2} \sum_{(k,l)\neq(i,j);(k,l)\neq(j,i)} s_{k,l}\sigma^{k,l}]m(\{s_{k,l}\})$$

where

$$C(\{\sigma^{k,l}\}) = c_1^{-1}[\det(\sigma^{k,l})]^{n/2}$$

$$c_1 = 2^{(Nn/2)} \times \pi^{N(N-1)/4} \times \Gamma(n/2)\Gamma((n-1)/2)\cdots\Gamma((n-N+1)/2)$$

$$m(\{s_{k,l}\}) = [\det(s_{k,l})]^{(n-N-2)/2}$$

According to (6.7), the UMPU test for hypothesis (6.4) has the form:

$$\varphi_{i,j}(\{s_{k,l}\}) = \begin{cases} 0, & if \; c'_{i,j}(\{s_{k,l}\}) < s_{i,j} < c''_{i,j}(\{s_{k,l}\}), \; (k,l) \neq (i,j), (j,i) \\ 1, & if \; s_{i,j} \leq c'_{i,j}(\{s_{k,l}\}) \; or \; s_{i,j} \geq c''_{i,j}(\{s_{k,l}\}), \; (k,l) \neq (i,j), (j,i), \end{cases}$$

$$\tag{6.10}$$

where the critical values $c'_{i,j}$, $c''_{i,j}$ are defined from the equations (according to (6.8) and (6.9))

$$\frac{\int_{I\cap[c'_{i,j};c''_{i,j}]}[\det(s_{k,l})]^{(n-N-2)/2}ds_{i,j}}{\int_I[\det(s_{k,l})]^{(n-N-2)/2}ds_{i,j}} = 1 - \alpha \tag{6.11}$$

$$\int_{I\cap(-\infty;c'_{i,j}]} s_{i,j}[\det(s_{k,l})]^{(n-N-2)/2}ds_{i,j} +$$
$$+ \int_{I\cap[c''_{i,j};+\infty)} s_{i,j}[\det(s_{k,l})]^{(n-N-2)/2}ds_{i,j} = \tag{6.12}$$
$$= \alpha \int_I s_{i,j}[\det(s_{k,l})]^{(n-N-2)/2}ds_{i,j}$$

where I is the interval of values of $s_{i,j}$ such that the matrix $S = (s_{k,l})$ is positive definite and α is the significance level of the test.

Let $S = (s_{k,l})$ be positive definite (this is true with probability 1 if $n > N$). Consider $\det(s_{k,l})$ as a function of the variable $s_{i,j}$ only, when fixing the values of all others $\{s_{k,l}\}$. This determinant is a quadratic polynomial of $s_{i,j}$:

$$\det(s_{k,l}) = -as_{i,j}^2 + bs_{i,j} + c \tag{6.13}$$

Let $K = (n - N - 2)/2$. Denote by x_1, x_2 $(x_1 < x_2)$ the roots of the equation $-ax^2 + bx + c = 0$. One has

$$\int_f^d (ax^2 - bx - c)^K dx = (-1)^K a^K (x_2 - x_1)^{2K+1} \int_{\frac{f-x_1}{x_2-x_1}}^{\frac{d-x_1}{x_2-x_1}} u^K (1-u)^K du$$

Therefore, the equation (6.11) takes the form:

$$\int_{\frac{c'-x_1}{x_2-x_1}}^{\frac{c''-x_1}{x_2-x_1}} u^K (1-u)^K \, du = (1-\alpha) \int_0^1 u^K (1-u)^K \, du \qquad (6.14)$$

or

$$\frac{\Gamma(2K+2)}{\Gamma(K+1)\Gamma(K+1)} \int_{\frac{c'-x_1}{x_2-x_1}}^{\frac{c''-x_1}{x_2-x_1}} u^K (1-u)^K \, du = (1-\alpha) \qquad (6.15)$$

It means that conditional distribution of $S_{i,j}$ when all other $S_{k,l}$ are fixed, $S_{k,l} = s_{k,l}$ is the beta distribution $Be(K+1, K+1)$.

Beta distribution $Be(K+1, K+1)$ is symmetric with respect to the point $\frac{1}{2}$. Therefore, the significance level condition (6.11) and unbiasedness condition (6.12) are satisfied if and only if:

$$\frac{c''-x_1}{x_2-x_1} = 1 - \frac{c'-x_1}{x_2-x_1}$$

Let q be the $\frac{\alpha}{2}$-quantile of beta distribution $Be(K+1, K+1)$, i.e., $F_{Be}(q) = \frac{\alpha}{2}$. Then thresholds c', c'' are defined by:

$$\begin{aligned} c' &= x_1 + (x_2 - x_1)q \\ c'' &= x_2 - (x_2 - x_1)q \end{aligned} \qquad (6.16)$$

Finally, the UMPU test for testing conditional independence of X_i, X_j has the form

$$\varphi_{i,j} = \begin{cases} 0, \ 2q - 1 < \dfrac{as_{i,j} - \frac{b}{2}}{\sqrt{\frac{b^2}{4} + ac}} < 1 - 2q \\ 1, \ \text{otherwise} \end{cases} \qquad (6.17)$$

where a, b, c are defined in (6.13).

6.3.5 Sample Partial Correlation Test

It is known [59] that hypothesis $\sigma^{i,j} = 0$ is equivalent to the hypothesis $\rho^{i,j} = 0$, where $\rho^{i,j}$ is the partial correlation between X_i and X_j given X_k, $k \in N(i, j) = \{1, 2, \ldots, N\} \setminus \{i, j\}$:

$$\rho^{i,j} = -\frac{\sigma^{i,j}}{\sqrt{\sigma^{i,i}\sigma^{j,j}}} = \frac{-\Sigma^{i,j}}{\sqrt{\Sigma^{i,i}\Sigma^{j,j}}}$$

where for a given matrix $A = (a_{k,l})$ we denote by $A^{i,j}$ the co-factor of the element $a_{i,j}$. Denote by $r^{i,j}$ sample partial correlation

$$r^{i,j} = \frac{-S^{i,j}}{\sqrt{S^{i,i}S^{j,j}}}$$

where $S^{i,j}$ is the cofactor of the element $s_{i,j}$ in the matrix S of sample covariances.

Well-known sample partial correlation test for testing hypothesis $\rho^{i,j} = 0$ has the form [1]:

$$\varphi_{i,j} = \begin{cases} 0, & |r^{i,j}| \le c_{i,j} \\ 1, & |r^{i,j}| > c_{i,j} \end{cases} \tag{6.18}$$

where $c_{i,j}$ is $(1 - \alpha/2)$-quantile of the distribution with the following density function

$$f(x) = \frac{1}{\sqrt{\pi}} \frac{\Gamma(n - N + 1)/2)}{\Gamma((n - N)/2)} (1 - x^2)^{(n-N-2)/2}, \quad -1 \le x \le 1$$

Note that in practical applications the following Fisher transformation is applied:

$$z_{i,j} = \frac{\sqrt{n}}{2} \ln\left(\frac{1 + r^{i,j}}{1 - r^{i,j}}\right)$$

Under condition $\rho^{i,j} = 0$ statistic $Z_{i,j}$ has asymptotically standard normal distribution. That is why the following test is largely used in applications [19], [20], [18]:

$$\varphi_{i,j} = \begin{cases} 0, & |z_{i,j}| \le c_{i,j} \\ 1, & |z_{i,j}| > c_{i,j} \end{cases} \tag{6.19}$$

where the constant $c_{i,j}$ is $(1 - \alpha/2)$-quantile of standard normal distribution.

The main result of this Section is the following theorem.

Theorem 6.1 *Sample partial correlation test* (6.18) *is equivalent to UMPU test* (6.17) *for testing hypothesis* $\rho^{i,j} = 0$ *vs* $\rho^{i,j} \ne 0$.

Proof it is sufficient to prove that

$$\frac{S^{i,j}}{\sqrt{S^{i,i}S^{j,j}}} = \frac{as_{i,j} - \frac{b}{2}}{\sqrt{\frac{b^2}{4} + ac}} \tag{6.20}$$

To verify this equation, we introduce some notations. Let $A = (a_{k,l})$ be an $(N \times N)$ symmetric matrix. Fix $i < j$, $i, j = 1, 2, \ldots, N$. Denote by $A(x)$ the matrix obtained from A by replacing the elements $a_{i,j}$ and $a_{j,i}$ by x. Denote by $A^{i,j}(x)$ the co-factor of the element (i, j) in the matrix $A(x)$. Then the following statement is true

Lemma 6.1 *One has* $[det A(x)]' = -2A^{i,j}(x)$.

Proof of the Lemma one has from the general Laplace decomposition of $\det A(x)$ by two rows i and j:

$$\det(A(x)) = \det \begin{pmatrix} a_{i,i} & x \\ x & a_{j,j} \end{pmatrix} A^{\{i,j\},\{i,j\}} + \sum_{k<j, k\neq i} \det \begin{pmatrix} a_{i,k} & x \\ a_{j,k} & a_{j,j} \end{pmatrix} A^{\{i,j\},\{k,j\}} +$$

$$+ \sum_{k>j} \det \begin{pmatrix} x & a_{i,k} \\ a_{j,j} & a_{j,k} \end{pmatrix} A^{\{i,j\},\{j,k\}} + \sum_{k<i} \det \begin{pmatrix} a_{i,k} & a_{i,i} \\ a_{j,k} & x \end{pmatrix} A^{\{i,j\},\{k,i\}} +$$

$$\sum_{k>i, k\neq j} \det \begin{pmatrix} a_{i,i} & a_{i,k} \\ x & a_{j,k} \end{pmatrix} A^{\{i,j\},\{i,k\}} + \sum_{k<l, k,l\neq i,j} \det \begin{pmatrix} a_{i,k} & a_{i,l} \\ a_{j,k} & a_{j,l} \end{pmatrix} A^{\{i,j\},\{k,l\}}$$

where $A^{\{i,j\},\{k,l\}}$ is the co-factor of the matrix $\begin{pmatrix} a_{i,k} & a_{i,l} \\ a_{j,k} & a_{j,l} \end{pmatrix}$ in the matrix A. Taking the derivative of $\det A(x)$, one gets

$$[\det(A(x))]' = -2x A^{\{i,j\},\{i,j\}} - \sum_{k<j, k\neq i} a_{j,k} A^{\{i,j\},\{k,j\}} + \sum_{k>j} a_{j,k} A^{\{i,j\},\{j,k\}} +$$

$$+ \sum_{k<i} a_{k,i} A^{\{i,j\},\{k,i\}} - \sum_{k>i, k\neq j} a_{k,i} A^{\{i,j\},\{k,i\}} = -2A^{i,j}(x)$$

The last equation follows from the symmetry conditions $a_{k,l} = a_{l,k}$ and from Laplace decompositions of $A^{i,j}(x)$ by the row j and the column i. Lemma is proved.

Now we come back to the proof of the theorem. One has $\det(S(x)) = -ax^2 + bx + c$, where a, b, c are the same as in (6.13). Therefore, by Lemma 6.1, one has $[\det S(x)]' = -2ax + b = -2S^{i,j}(x)$, i.e., $S^{i,j}(x) = ax - b/2$. Let $x = s_{i,j}$ then $as_{i,j} - \frac{b}{2} = S^{i,j}$. To prove the theorem, it is sufficient to prove that $\sqrt{S^{i,i} S^{j,j}} = \sqrt{\frac{b^2}{4} + ac}$. Let $x_2 = \frac{b+\sqrt{b^2+4ac}}{2a}$ be the maximum root of equation $ax^2 - bx - c = 0$. Then $ax_2 - \frac{b}{2} = \sqrt{\frac{b^2}{4} + ac}$. Consider

$$r^{i,j}(x) = \frac{-S^{i,j}(x)}{\sqrt{S^{i,i} S^{j,j}}}$$

According to Silvester determinant identity, one can write:

$$S^{\{i,j\},\{i,j\}} \det S(x) = S^{i,i} S^{j,j} - [S^{i,j}(x)]^2$$

Therefore, for $x = x_1$ and $x = x_2$, one has

$$S^{i,i} S^{j,j} - [S^{i,j}(x)]^2 = 0$$

That is for $x = x_1$ and $x = x_2$ one has $r^{i,j}(x) = \pm 1$. The equation $S^{i,j}(x) = ax - \frac{b}{2}$ implies that when x is increasing from x_1 to x_2 then $r^{i,j}(x)$ is decreasing from 1 to -1. That is $r^{i,j}(x_2) = -1$, i.e., $ax_2 - \frac{b}{2} = \sqrt{S^{i,i} S^{j,j}}$. Therefore,

$$\sqrt{S^{i,i} S^{j,j}} = \sqrt{\frac{b^2}{4} + ac}$$

The Theorem is proved.

Finally, the UMPU test for testing conditional independence of X_i and X_j can be written in the following form

$$\varphi_{i,j}^{umpu} = \begin{cases} 0, & 2q - 1 < r^{i,j} < 1 - 2q \\ 1 & \text{otherwise} \end{cases} \tag{6.21}$$

where $r^{i,j}$ is the sample partial correlation, and q is the $\frac{\alpha}{2}$-quantile of beta distribution $Be(\frac{n-N}{2}, \frac{n-N}{2})$.

6.3.6 Optimal Multiple Decision Procedures

Define the multiple decision procedure $\delta^{ou} = (\varphi_{i,j}^{umpu})_{i,j=1,...,N}$. The following theorem shows that δ^{ou} is optimal in the class of w-unbiased multiple decision procedures if significance levels of individual tests are related to individual losses.

Theorem 6.2 *Let the loss function w be additive (see 4.6) and the significance levels of individual tests are related with individual losses by*

$$\alpha_{i,j} = \frac{b_{i,j}}{a_{i,j} + b_{i,j}}, \quad i \neq j, \; i, j = 1, 2, \ldots, N. \tag{6.22}$$

Then the procedure δ^{ou} is optimal multiple decision procedure for Gaussian graphical model selection in the class of w-unbiased procedures.

Proof Let $\delta = (\varphi_{i,j})_{i,j=1,...,N}$. If the loss function w satisfies (4.6) then the risk of procedure δ is the sum of the risks of individual tests $\varphi_{i,j}$. Indeed, one has

$$\text{Risk}(S, \theta; \delta) = \sum_{Q \in \mathcal{G}} w(S, Q) P_\theta(\delta(x) = d_Q) =$$

$$= \sum_{Q \in \mathcal{G}} [\sum_{\{i,j : s_{i,j} = 0; q_{i,j} = 1\}} a_{i,j} + \sum_{\{i,j : s_{i,j} = 1; q_{i,j} = 0\}} b_{i,j}] P_\theta(\delta(x) = d_Q) =$$

$$= \sum_{i,j=1, s_{i,j}=0}^{N} a_{i,j} \sum_{Q, q_{i,j}=1} P_\theta(\delta(x) = d_Q) + \sum_{i,j=1, s_{i,j}=1}^{N} b_{i,j} \sum_{Q, q_{i,j}=0} P_\theta(\delta(x) = d_Q) =$$

$$= \sum_{i,j=1, s_{i,j}=0}^{N} a_{i,j} P_\theta(\varphi_{i,j}(x) = 1) + \sum_{i,j=1, s_{i,j}=1}^{N} b_{i,j} P_\theta(\varphi_{i,j}(x) = 0) =$$

$$= \sum_{i=1}^{N} \sum_{j=1}^{N} \text{Risk}(s_{i,j}, \theta; \varphi_{i,j})$$

where

$$\text{Risk}(s_{i,j}, \theta; \varphi_{i,j}) = \begin{cases} a_{i,j} P_\theta(\varphi_{i,j} = 1), & \text{if } \theta \in \omega_{i,j} \\ b_{i,j} P_\theta(\varphi_{i,j} = 0), & \text{if } \theta \in \omega_{i,j}^{-1} \end{cases}$$

Now we prove that δ^{ou} is a w-unbiased multiple decision procedure. One has from relation (6.22) and unbiasedness (see Sect. 6.2) of φ^{umpu}

$$a_{i,j} P_\theta(\varphi^{\text{umpu}} i, j = 1) \le b_{i,j} P_\theta(\varphi_{i,j}^{\text{umpu}} = 0), \quad \text{if } \theta \in \omega_{i,j},$$

and

$$b_{i,j} P_\theta(\varphi^{\text{umpu}} i, j = 0) \le a_{i,j} P_\theta(\varphi_{i,j}^{\text{umpu}} = 1), \quad \text{if } \theta \in \omega_{i,j}^{-1}$$

It implies that

$$\text{Risk}(s_{i,j}, \theta; \varphi_{i,j}^{\text{umpu}}) \le \text{Risk}(s_{i,j}', \theta; \varphi_{i,j}^{\text{umpu}}), \quad \forall s_{i,j}, s_{i,j}' = 0, 1$$

Therefore,

$$\text{Risk}(S, \theta; \delta^{ou}) = \sum_{i=1}^{N} \sum_{j=1}^{N} \text{Risk}(s_{i,j}, \theta; \varphi_{i,j}^{\text{umpu}}) \le$$

$$\leq \sum_{i=1}^{N} \sum_{j=1}^{N} \text{Risk}(s'_{i,j}, \theta; \varphi_{i,j}^{\text{umpu}}) = \text{Risk}(S', \theta; \delta^{ou}),$$

for any $S, S' \in \mathscr{G}, \theta \in \Omega_S$.

Finally, we prove that δ^{ou} is optimal in the class of w-unbiased procedures. Let $\delta = (\varphi_{i,j})_{i,j=1,\dots,N}$ be any w-unbiased procedure. One has:

$$\text{Risk}(S, \theta; \delta) \leq \text{Risk}(S', \theta; \delta), \quad \forall S, S' \in \mathscr{G},$$

$\theta \in \Omega_S$. Take S' such that S' and S differ only in two positions (i, j) and (j, i). In this case, one has

$$\text{Risk}(S, \theta; \delta) = 2\text{Risk}(s_{i,j}, \theta; \varphi_{i,j}) + \sum_{(k,l) \neq (i,j),(j,i)} \text{Risk}(s_{k,l}, \theta; \varphi_{k,l})$$

and:

$$\text{Risk}(S', \theta; \delta) = 2\text{Risk}(s'_{i,j}, \theta; \varphi_{i,j}) + \sum_{(k,l) \neq (i,j),(j,i)} \text{Risk}(s_{k,l}, \theta; \varphi_{k,l})$$

Therefore,

$$\text{Risk}(s_{i,j}, \theta; \varphi_{i,j}) \leq \text{Risk}(s'_{i,j}, \theta; \varphi_{i,j})$$

It implies unbiasedness of the test $\varphi_{i,j}$, i.e.:

$$P_\theta(\varphi_{i,j} = 1) \leq \frac{b_{i,j}}{a_{i,j} + b_{i,j}} = \alpha_{i,j}, \quad \theta \in \omega_{i,j}$$

and:

$$P_\theta(\varphi_{i,j} = 1) \geq \frac{b_{i,j}}{a_{i,j} + b_{i,j}} = \alpha_{i,j}, \quad \theta \in \omega_{i,j}^{-1}$$

One has

$$P_\theta(\varphi_{i,j} = 1) = \int_{\varphi_{i,j}=1} f(x, \theta) dx,$$

where $\theta = (\mu, \Sigma)$, and

$$f(x; \theta) = \frac{1}{(2\pi)^{N/2} \det(\Sigma)^{1/2}} \exp(-\frac{1}{2}(x - \mu)^T \Sigma^{-1}(x - \mu))$$

Therefore, the function $P_\theta(\varphi_{i,j} = 1)$ is continuous with respect to $\theta \in \Omega$. One has $\omega_{i,j} = \{\theta : \sigma^{i,j} = 0\}$, $\omega_{i,j}^{-1} = \Omega \setminus \omega_{i,j}$. Unbiasedness of the test $\varphi_{i,j}$ and continuity of the function $P_\theta(\varphi_{i,j} = 1)$ imply $P_\theta(\varphi_{i,j} = 1) = \alpha_{i,j}$, for $\theta \in \omega_{i,j}$. Thus, the individual test $\varphi_{i,j}$ has the significance level $\alpha_{i,j} = b_{i,j}/(a_{i,j} + b_{i,j})$. Taking into account that $\varphi_{i,j}^{umpu}$ is UMP in the class of unbiased tests with significance level $\alpha_{i,j}$, one gets

$$\text{Risk}(s_{i,j}, \theta; \varphi_{i,j}^{umpu}) = a_{i,j}\alpha_{i,j} = \text{Risk}(s_{i,j}, \theta; \varphi_{i,j}), \ \theta \in \omega_{i,j}$$

and

$$\text{Risk}(s_{i,j}, \theta; \varphi_{i,j}^{umpu}) = b_{i,j} P_\theta(\varphi_{i,j}^{umpu} = 0) \le b_{i,j} P_\theta(\varphi_{i,j} = 0) = \text{Risk}(s_{i,j}, \theta; \varphi_{i,j}),$$

for $\theta \in \omega_{i,j}^{-1}$. Therefore,

$$\text{Risk}(S, \theta; \delta^{ou}) = \sum_{i=1}^{N}\sum_{j=1}^{N}\text{Risk}(s_{i,j}, \theta; \varphi_{i,j}^{umpu}) \le$$

$$\le \sum_{i=1}^{N}\sum_{j=1}^{N}\text{Risk}(s_{i,j}, \theta; \varphi_{i,j}) = \text{Risk}(S, \theta; \delta),$$

for any w-unbiased procedure δ. The theorem is proved.

Remark If all individual hypotheses have the same importance, then one can set $a_{i,j} = a$, $b_{i,j} = b$. In this case, the following is true. Let $0 < \alpha < 1$, and the loss function w be defined by (4.6) with $a_{i,j} = 1 - \alpha$, $b_{i,j} = \alpha$, $i, j = 1, 2, \ldots, N$. Then for any w-unbiased multiple decision procedure δ, one has

$$(1-\alpha)E_\theta[Y_I(S, \delta^{ou})]+\alpha E_\theta[Y_{II}(S, \delta^{ou})] \le (1-\alpha)E_\theta[Y_I(S, \delta)]+\alpha E_\theta[Y_{II}(S, \delta)],$$

for any $S \in \mathcal{G}, \theta \in \Omega_S$. Indeed, for any procedure δ, one has

$$\text{Risk}(S, \theta; \delta) = (1 - \alpha)E_\theta[Y_I(S, \delta)] + \alpha E_\theta[Y_{II}(S, \delta)]$$

The theorem implies

$$\text{Risk}(S, \theta; \delta^{ou}) \le \text{Risk}(S, \theta; \delta)$$

for any w-unbiased procedure δ. Note that in this case, α is the significance level common to all the individual tests.

6.4 Optimal Threshold Graph Identification in Pearson Correlation Network

In this section, we investigate the properties of statistical procedures for threshold graph (TG) identification in Gaussian Pearson correlation network. In this case, the parameter θ can be defined by $\theta = (\mu, \Sigma)$, where $\mu \in R^N$, and Σ is a positive definite matrix ($N \times N$). True threshold graph for a given threshold ρ_0 is defined as follows: graph has N nodes, edge (i, j) is included in the true threshold graph if $\rho(X_i, X_j) > \rho_0$. Here $\rho(X_i, X_j)$ is the Pearson correlation between X_i and X_j. To identify the true threshold graph from observations, one can test the individual hypotheses:

$$h_{i,j} : \rho_{i,j} \leq \rho_0 \text{ vs } k_{i,j} : \rho_{i,j} > \rho_0, \quad i \neq j, i, j = 1, 2, \ldots, N \quad (6.23)$$

Hypothesis $h_{i,j}$ means that there is no edge between vertices i and j in the true TG. Alternative $k_{i,j}$ means that there is an edge between vertices i and j in the true TG. Let $\varphi_{i,j}(x)$ be a test for individual hypothesis $h_{i,j}$: $\varphi_{i,j} = 1$ means that we reject the hypothesis, and $\varphi_{i,j} = 0$ means that we accept the hypothesis. Then the associated multiple decision statistical procedure for TG identification is defined as

$$\delta(x) = d_Q \text{ if } \Phi(x) = Q \quad (6.24)$$

where $\Phi(x)$ is the following matrix

$$\Phi(x) = \begin{pmatrix} 0 & \varphi_{1,2}(x) & \ldots & \varphi_{1,N}(x) \\ \varphi_{2,1}(x) & 0 & \ldots & \varphi_{2,N}(x) \\ \ldots & \ldots & \ldots \ldots \\ \varphi_{N,1}(x) & \varphi_{N,2}(x) & \ldots & 0 \end{pmatrix} \quad (6.25)$$

Let

$$x(t) = (x_1(t), x_2(t), \ldots, x_N(t)), \quad t = 1, 2, \ldots, n \quad (6.26)$$

be a sample of observations from random vector $X = (X_1, X_2, \ldots, X_N)$. Let $G = \{g_{c,d} : c \in R, d \in R^{N \times n}\}$ be the group of scale/shift transformations of the sample space $R^{N \times n}$ where for any $c \in R, d \in R^{N \times n}$ the transformation $g_{c,d} : R^{N \times n} \rightarrow R^{N \times n}$ is defined as $g_{c,d} = cx + d, x \in R^{N \times n}$. Function $M(x)$ is called invariant with respect to the group G if $M(g_{c,d}x) = M(x), \forall g_{c,d} \in G, \forall x \in R^{N \times n}$. Similarly, statistical test φ is called invariant with respect to the group G if $\varphi(g_{c,d}x) = \varphi(x)$ for $\forall g_{c,d} \in G$ Function $M(x)$ is maximal invariant with respect to the group G if $M(x_1) = M(x_2)$ implies $x_1 = g_{c,d}x_2$ for some $g_{c,d} \in G$. As follows from Neyman-Pearson fundamental lemma ([62], pp. 59) that if maximal invariant has monotone likelihood ratio then test based on maximal invariant is UMP invariant. It is proved in [62] that sample correlation $r_{i,j}$ is maximal invariant with respect to the group G. Distribution of $r = r_{i,j}$ depends on $\rho = \rho_{i,j}$ only and can be written as [1]:

$$f_\rho(r) = \frac{n-2}{\sqrt{2\pi}} \frac{\Gamma(n-1)}{\Gamma(n-1/2)} (1 - \rho^2)^{1/2(n-1)} (1 - r^2)^{1/2(n-4)} (1 - \rho r)^{-n+3/2} \times$$

$$\times F(\frac{1}{2}; \frac{1}{2}; n - \frac{1}{2}; \frac{1 + \rho r}{2})$$

where

$$F(a, b, c, x) = \sum_{j=0}^{\infty} \frac{\Gamma(a+j)}{\Gamma(a)} \frac{\Gamma(b+j)}{\Gamma(b)} \frac{\Gamma(c)}{\Gamma(c+j)} \frac{x^j}{j!}$$

$$(6.27)$$

The distribution (6.27) has monotone likelihood ratio [62]. Therefore, the following test:

$$\varphi_{i,j}^{\mathrm{Corr}}(x_i, x_j) = \begin{cases} 1, & \dfrac{r_{i,j} - \rho_0}{\sqrt{1 - r_{i,j}^2}} > c_{i,j} \\ \\ 0, & \dfrac{r_{i,j} - \rho_0}{\sqrt{1 - r_{i,j}^2}} \le c_{i,j} \end{cases}$$

$$(6.28)$$

is the uniformly most powerful invariant test with respect to the group G for testing the individual hypothesis $h_{i,j}$ defined by (6.23). Here $c_{i,j}$ is chosen to make the significance level of the test equal to prescribed value $\alpha_{i,j}$. This means that for any test $\varphi'_{i,j}$ invariant with respect to the group G with $E_{\rho_0}\varphi'_{i,j} = \alpha_{i,j}$ one has

$$P_{\rho_{i,j}}(\varphi_{i,j}^{\mathrm{Corr}} = 1) \ge P_{\rho_{i,j}}(\varphi'_{i,j} = 1), \quad \rho_{i,j} > \rho_0$$
$$P_{\rho_{i,j}}(\varphi_{i,j}^{\mathrm{Corr}} = 1) \le P_{\rho_{i,j}}(\varphi'_{i,j} = 1), \quad \rho_{i,j} \le \rho_0$$

$$(6.29)$$

where $P_{\rho_{i,j}}(\varphi_{i,j} = 1)$ is the probability of the rejection of hypothesis $h_{i,j}$ for a given $\rho_{i,j}$.

The statistical procedure $\delta(x)$ is called invariant with respect to the group G if $\delta(g_{c,d}x) = \delta(x)$ for $\forall g_{c,d} \in G$ We consider the following class $\mathscr{D}$ of statistical procedures $\delta(x)$ for network structures identification:

1. Any statistical procedure $\delta(x) \in \mathscr{D}$ is invariant with respect to the group G of shift/scale transformations of the sample space.
2. Risk function of any statistical procedure $\delta(x) \in \mathscr{D}$ is continuous with respect to parameter.
3. Individual tests $\varphi_{i,j}$ generated by any $\delta(x) \in \mathscr{D}$ depend on observations $x_i(t), x_j(t), t = 1, \ldots, n$ only.

Define statistical procedure $\delta_{\mathrm{Corr}}(x)$ by

$$\delta_{\text{Corr}}(x) = d_Q \text{ if } \Phi^{\text{Corr}}(x) = Q \tag{6.30}$$

where

$$\Phi^{\text{Corr}}(x) = \begin{pmatrix} 0, & \varphi_{1,2}^{\text{Corr}}(x), & \ldots, & \varphi_{1,N}^{\text{Corr}}(x) \\ \varphi_{2,1}^{\text{Corr}}(x), & 0, & \ldots, & \varphi_{2,N}^{\text{Corr}}(x) \\ \ldots & \ldots & \ldots & \ldots \\ \varphi_{N,1}^{\text{Corr}}(x), & \varphi_{N,2}^{\text{Corr}}(x), & \ldots, & 0 \end{pmatrix} \tag{6.31}$$

The main result of this section is given in the following theorem

Theorem 6.3 *Let* (X, γ^P) *be a Gaussian Pearson correlation network, loss function be additive, and losses* $a_{i,j}, b_{i,j}$ *of individual tests* $\varphi_{i,j}$ *for testing hypotheses* $h_{i,j}$ *are connected with the significance levels* $\alpha_{i,j}$ *of the tests by*

$$a_{i,j} = 1 - \alpha_{i,j}, \quad b_{i,j} = \alpha_{i,j}.$$

Then the statistical procedure δ_{Corr} *defined by (6.30) and (6.31) is optimal multiple decision statistical procedure in the class* $\mathscr{D}$.

Proof Statistical tests φ^{Corr} defined by (6.28) are invariant with respect to the group G of scale/shift transformation of the sample space. It implies that multiple decision statistical procedure δ_{Corr} is invariant with respect to the group G. Statistical tests φ^{Corr} depend on $x_i(t), x_j(t), t = 1, \ldots, n$ only. According to [55, 60], risk function of statistical procedure δ_{Corr} for additive loss function can be written as

$$R(S, (\rho_{i,j}), \delta_{\text{Corr}}) = \sum_{i,j:i\neq j} r(\rho_{i,j}, \varphi_{i,j}^{\text{Corr}}) \tag{6.32}$$

where $r(\rho_{i,j}, \varphi_{i,j}^{\text{Corr}})$ is the risk function of the individual test $\varphi_{i,j}^{\text{Corr}}$ for the individual hypothesis $h_{i,j}$, $(\rho_{i,j})$ is the matrix of correlations. One has

$$r(\rho_{i,j}, \varphi_{i,j}^{\text{Corr}}) = \begin{cases} (1 - \alpha_{i,j}) E_{\rho_{i,j}} \varphi_{i,j}^{\text{Corr}}, & \text{if } \rho_{i,j} \leq \rho_0 \\ \\ \alpha_{i,j}(1 - E_{\rho_{i,j}} \varphi_{i,j}^{\text{Corr}}), & \text{if } \rho_{i,j} > \rho_0 \end{cases}$$

Since $E_{\rho_{i,j}} \varphi_{i,j}^{\text{Corr}} = \alpha_{i,j}$ if $\rho_{i,j} = \rho_0$ then function $r(\rho_{i,j}, \varphi_{i,j}^{\text{Corr}})$ is continuous as function of $\rho_{i,j}$. Therefore, multiple decision statistical procedure δ_{Corr} belongs to the class $\mathscr{D}$.

Let $\delta' \in \mathscr{D}$ be another statistical procedure for TG identification. Then $\varphi'_{i,j}(x)$ depends on $x_i = (x_i(1), \ldots, x_i(n)), x_j = (x_j(1), \ldots, x_j(n))$ only. Statistical procedure $\delta' \in \mathscr{D}$ is invariant with respect to the group G if and only if associated individual tests $\varphi'_{i,j}(x)$ are invariant with respect to the group G for all $i, j = 1, \ldots, N, i \neq j$. One has for an additive loss function (see [55, 60]):

$$R(S, \theta, \delta') = \sum_{i,j} r(\theta, \varphi'_{i,j}) \tag{6.33}$$

where

$$r(\theta, \varphi'_{i,j}) = \begin{cases} (1 - \alpha_{i,j}) E_\theta \varphi'_{i,j}, & \text{if } \rho_{i,j} \leq \rho_0 \\ \alpha_{i,j}(1 - E_\theta \varphi'_{i,j}), & \text{if } \rho_{i,j} > \rho_0 \end{cases}$$

Since tests $\varphi'_{i,j}$ are invariant with respect to the group G then the distributions of tests $\varphi'_{i,j}$ depend on $\rho_{i,j}$ only [62]. It implies

$$r(\theta, \varphi'_{i,j}) = r(\rho_{i,j}, \varphi'_{i,j})$$

and

$$R(S, \theta, \delta') = R(S, (\rho_{i,j}), \delta')$$

Risk functions of statistical procedures from the class $\mathscr{D}$ are continuous with respect to parameter, and one gets $E_{\rho_0} \varphi'_{i,j} = \alpha_{i,j}$. It means that the test $\varphi'_{i,j}$ has significance levels $\alpha_{i,j}$. The test $\varphi^{\text{Corr}}_{i,j}$ is UMP invariant test of the significance level $\alpha_{i,j}$, therefore one has

$$r(\rho_{i,j}, \varphi^{\text{Corr}}_{i,j}) \leq r(\rho_{i,j}, \varphi'_{i,j}), \quad i, j = 1, \ldots, N.$$

Then

$$R(S, (\rho_{i,j}), \delta_{\text{Corr}}) \leq R(S, (\rho_{i,j}), \delta'), \quad \forall S \in \mathscr{G}, \ \forall \delta' \in \mathscr{D}, \ \forall \rho$$

The theorem is proved.

Note that in many publications on market network analysis the edge (i, j) is included in the market graph if $r_{i,j} > \rho_0$, and it is not included in the market graph if $r_{i,j} \leq \rho_0$. This statistical procedure corresponds to the statistical procedure $\delta_{\text{Corr}}(x)$ with $\alpha_{i,j} = 0.5, i, j = 1, 2, \ldots, N, i \neq j$. Therefore, this procedure is optimal in the class $\mathscr{D}$ for the risk function equals to the sum of the expected numbers of false edge inclusions and of the expected number of false edge exclusions.

For the case $\rho_0 = 0$, the statistical procedure $\delta_{\text{Corr}}(x)$ is in addition optimal in the class of unbiased statistical procedures for the same additive loss function. This follows from optimality of the tests $\varphi^{\text{Corr}}(x)$ in the class of unbiased tests and some results from [54, 60].

The class $\mathscr{D}$ is defined by three conditions. All of them are important in the proof of the Theorem 6.3 and cannot be removed. The condition that risk function is continuous with respect to parameter cannot be removed, because when we consider the class $\mathscr{D}$ without this condition the statistical procedure δ_{Corr} with significance levels of individual tests $\alpha_{i,j}$ is no more optimal in this large class. A counterexample is given by any statistical procedure of the same type δ_{Corr}, but with different significance levels of individual tests. Note that all these statistical procedures have a discontinuous risk function for the losses $a_{i,j} = 1 - \alpha_{i,j}$, $b_{i,j} = \alpha_{i,j}$. The condition that individual tests $\varphi_{i,j}$ depend on observations $x_i(t)$, $x_j(t)$ only also cannot be removed. A counterexample is given by Holm type step down procedures that can be shown by numerical experiments.

6.5 Optimal Threshold Graph Identification in Sign Similarity Network

In this section, we study optimal statistical procedures for threshold graph identification in a sign similarity (sign correlation based) network. Our construction of optimal procedures is based on the simultaneous inference of optimal two-decision procedures.

In the sign similarity network model, the weight of the edge (i, j) is defined by

$$\gamma_{i,j}^{Sg} = p^{i,j} = P((X_i - E(X_i))(X_j - E(X_j)) > 0) \tag{6.34}$$

For a given threshold $\gamma_0^{Sg} = p_0$, the true threshold graph in the sign similarity network model is constructed as follows: The edge between two nodes i and j is included in the true threshold graph if $p^{i,j} > p_0$, where $p^{i,j}$ is the probability of the sign coincidence of the random variables associated with nodes i and j.

In practice we are given a sample of observations $x(1), x(2), \ldots, x(n)$ from X, which are modeled by a family of random vectors

$$X(t) = (X_1(t), X_2(t), \ldots, X_N(t)), \quad t = 1, 2, \ldots, n$$

where n is the number of observations (sample size) and vectors $X(t)$ are independent and identically distributed as $X = (X_1, X_2, \ldots, X_N)$. Henceforth we assume that expectations $E(X_i), i = 1, 2, \ldots, N$ are known. We put (for simplicity) $E(X_i) = 0, i = 1, 2, \ldots, N$. In this case

$$p^{i,j} = P(X_i X_j > 0), \quad i, j = 1, 2, \ldots, N \tag{6.35}$$

Define the $N \times n$ matrix $x = (x_i(t))$. Consider the set $\mathscr{G}$ of all $N \times N$ symmetric matrices $G = (g_{i,j})$ with $g_{i,j} \in \{0, 1\}$, $i, j = 1, 2, \ldots, N$, $g_{i,i} = 0$, $i = 1, 2, \ldots, N$.

Any individual edge test can be reduced to a hypothesis testing problem:

$$h_{i,j} : p^{i,j} \le p_0 \quad vs \quad k_{i,j} : p^{i,j} > p_0 \tag{6.36}$$

Let $\varphi_{i,j}(x)$ be a test for individual hypothesis (6.36). More precisely, $\varphi_{i,j}(x) = 1$ means that the hypothesis $h_{i,j}$ is rejected (the edge (i, j) is included in the threshold graph), $\varphi_{i,j}(x) = 0$ means that $h_{i,j}$ is accepted (the edge (i, j) is not included in the threshold graph). Let $\Phi(x)$ be the matrix $\Phi(x) = (\varphi_{i,j}(x))_{i,j=1,...,N}$. Multiple decision statistical procedure $\delta(x)$ based on the simultaneous inference of individual edge tests (6.36) can be written as

$$\delta(x) = d_G, \text{ if } \Phi(x) = G \tag{6.37}$$

Let

$$p_{0,0}^{i,j} = P(X_i \leq 0, X_j \leq 0), \quad p_{1,1}^{i,j} = P(X_i > 0, X_j > 0)$$

$$p_{1,0}^{i,j} = P(X_i > 0, X_j \leq 0), \quad p_{0,1}^{i,j} = P(X_i \leq 0, X_j > 0)$$

One has $p^{i,j} = p_{0,0}^{i,j} + p_{1,1}^{i,j}$. Define

$$u_k(t) = \begin{cases} 0, & x_k(t) \leq 0 \\ 1, & x_k(t) > 0 \end{cases}$$

$k = 1, 2, \ldots, N$. Let us introduce the statistics

$$T_{1,1}^{i,j} = \sum_{t=1}^n u_i(t)u_j(t); \quad T_{0,0}^{i,j} = \sum_{t=1}^n (1 - u_i(t))(1 - u_j(t));$$
$$T_{0,1}^{i,j} = \sum_{t=1}^n (1 - u_i(t))u_j(t); \quad T_{1,0}^{i,j} = \sum_{t=1}^n u_i(t)(1 - u_j(t)); \tag{6.38}$$
$$V_{i,j} = T_{1,1}^{i,j} + T_{0,0}^{i,j}$$

To construct a multiple decision procedure, we use the following individual edge tests:

$$\varphi_{i,j}^{Sg}(x_i, x_j) = \begin{cases} 0, & V_{i,j} \leq c_{i,j} \\ 1, & V_{i,j} > c_{i,j} \end{cases} \tag{6.39}$$

where for a given significance level $\alpha_{i,j}$, the constant $c_{i,j}$ is defined as the smallest integer such that:

$$\sum_{k=c_{i,j}}^n \frac{n!}{k!(n-k)!}(p_0)^k(1-p_0)^{n-k} \leq \alpha_{i,j} \tag{6.40}$$

Let $\Phi^{Sg}(x)$ be the matrix

$$\Phi^{Sg}(x) = \begin{pmatrix} 1, & \varphi_{1,2}^{Sg}(x), & \ldots, & \varphi_{1,N}^{Sg}(x) \\ \varphi_{2,1}^{Sg}(x), & 1, & \ldots, & \varphi_{2,N}^{Sg}(x) \\ \ldots & \ldots & \ldots & \ldots \\ \varphi_{N,1}^{Sg}(x), & \varphi_{N,2}^{Sg}(x), & \ldots, & 1 \end{pmatrix}, \tag{6.41}$$

where $\varphi_{i,j}^{Sg}(x)$ is defined by (6.39) and (6.40). Now we can define our multiple decision statistical procedure for the threshold graph identification in sign similarity network by

$$\delta^{Sg}(x) = d_G, \text{ if } \Phi^{Sg}(x) = G \qquad (6.42)$$

The main result of this section is the following theorem.

Theorem 6.4 *Let the loss function w be additive, let the individual test $\varphi_{i,j}$ depend only on $u_i(t), u_j(t)$ for any i, j, and let the following symmetry conditions be satisfied*

$$p_{11}^{i,j} = p_{00}^{i,j}, \quad p_{10}^{i,j} = p_{01}^{i,j}, \quad \forall i, j = 1, 2, \ldots, N \qquad (6.43)$$

Then, for the statistical procedure δ^{Sg} defined by (6.39), (6.40), (6.41), and (6.42) for the threshold graph identification in the sign similarity network, one has

$$\text{Risk}(S, \delta^{Sg}) \leq \text{Risk}(S, \delta)$$

for any adjacency matrix S and any w-unbiased statistical procedure δ.

Proof We prove optimality in three steps. First, we prove that under the symmetry conditions (6.43), each individual test (6.39) is uniformly most powerful (UMP) in the class of tests based on $u_i(t), u_j(t)$ only for the individual hypothesis testing

$$h_{i,j} : p^{i,j} \leq p_0 \text{ vs } k_{i,j} : p^{i,j} > p_0 \qquad (6.44)$$

By symmetry the individual hypothesis (6.44) can be written as:

$$h_{i,j} : p_{00}^{i,j} \leq \frac{p_0}{2} \text{ vs } k_{i,j} : p_{00}^{i,j} > \frac{p_0}{2} \qquad (6.45)$$

Let for simplicity of notations $p_{0,0} = p_{0,0}^{i,j}; \; p_{0,1} = p_{0,1}^{i,j}; \; p_{1,0} = p_{1,0}^{i,j}; \; p_{1,1} = p_{1,1}^{i,j}, \; T_{0,0} = T_{0,0}^{i,j}; \; T_{0,1} = T_{0,1}^{i,j}; \; T_{1,0} = T_{1,0}^{i,j}; \; T_{1,1} = T_{1,1}^{i,j}.$ One has

$$T_{1,1} + T_{1,0} + T_{0,1} + T_{0,0} = n;$$

Symmetry implies

$$p_{0,0} + p_{1,0} = \frac{1}{2}$$

Let $t_{1,1}, t_{1,0}, t_{0,1}, t_{0,0}$ be nonnegative integers with $t_{1,1} + t_{1,0} + t_{0,1} + t_{0,0} = n$ and $C = n!/(t_{1,1}!t_{1,0}!t_{0,1}!t_{0,0}!)$. One has

$$P(T_{1,1} = t_{1,1}; T_{1,0} = t_{1,0}; T_{0,1} = t_{0,1}; T_{0,0} = t_{0,0}) = C p_{1,1}^{t_{1,1}} p_{1,0}^{t_{1,0}} p_{0,1}^{t_{0,1}} p_{0,0}^{t_{0,0}} =$$

$$= C p_{0,0}^{t_{1,1}+t_{0,0}} p_{1,0}^{t_{1,0}+t_{0,1}} = C_1 \exp\{(t_{1,1} + t_{0,0}) \ln \frac{p_{0,0}}{1/2 - p_{0,0}}\}$$

where $C_1 = C(1/2 - p_{0,0})^n$.

Then, the hypotheses (6.45) are equivalent to the hypotheses:

$$h'_{i,j} : \ln(\frac{p_{0,0}}{1/2 - p_{0,0}}) \le \ln(\frac{p_0}{1 - p_0}) \text{ vs } k'_{i,j} : \ln(\frac{p_{0,0}}{1/2 - p_{0,0}}) > \ln(\frac{p_0}{1 - p_0})$$
$$(6.46)$$

For $p_{0,0} = p_0/2$, the random variable $V = T_{1,1} + T_{0,0}$ has the binomial distribution $B(n, p_0)$. Therefore, the critical value for the test (6.39) is defined from (6.40). According to ([62], Ch.3, corollary 3.4.1) the test (6.39) is uniformly most powerful (UMP) at the level $\alpha_{i,j}$ for hypothesis testing (6.46).

Second, we prove that the statistical procedure (6.42) is w-unbiased. For any two-decision test for hypothesis testing (6.44), the risk function can be written as:

$$\text{Risk} = R(s_{i,j}, \varphi_{i,j}) = \begin{cases} a_{i,j} P(\varphi_{i,j}(x) = 1/p^{i,j}), & \text{if } s_{i,j} = 0(p^{i,j} \le p_0) \\ b_{i,j} P(\varphi_{i,j}(x) = 0/p^{i,j}), & \text{if } s_{i,j} = 1(p^{i,j} > p_0) \end{cases}$$

One has

$$a_{i,j} P(\varphi_{i,j}^{Sg}(x) = 1/p^{i,j}) \le b_{i,j} P(\varphi_{i,j}^{Sg}(x) = 0/p^{i,j}) \text{ if } p^{i,j} \le p_0$$

$$a_{i,j} P(\varphi_{i,j}^{Sg}(x) = 1/p^{i,j}) \ge b_{i,j} P(\varphi_{i,j}^{Sg}(x) = 0/p^{i,j}), \text{ if } p^{i,j} > p_0$$

which is equivalent to

$$R(s_{i,j}, \varphi_{i,j}^{Sg}) \le R(s'_{i,j}, \varphi_{i,j}^{Sg}), \forall s_{i,j}, s'_{i,j} \tag{6.47}$$

This relation implies

$$P(\varphi_{i,j}^{Sg}(x) = 1/p^{i,j} = p_0) = \alpha_{i,j} = \frac{b_{i,j}}{a_{i,j} + b_{i,j}}$$

This implies that the test $\varphi_{i,j}^{Sg}$ has significance level $\alpha_{i,j} = b_{i,j}/(a_{i,j} + b_{i,j})$. For the loss function (4.6) and any multiple decision statistical procedure δ, one has

$$R(H_S, \delta) = \sum_{Q \in \mathcal{G}} (\sum_{i,j:s_{i,j}=0;q_{i,j}=1} a_{i,j} + \sum_{i,j:s_{i,j}=1;q_{i,j}=0} b_{i,j}) P(x \in D_Q/H_S) =$$
$$= \sum_{s_{i,j}=0} a_{i,j} P(\varphi_{i,j}(x) = 1)/H_S) + \sum_{s_{i,j}=1} b_{i,j} P(\varphi_{i,j}(x) = 0)/H_S)$$
$$(6.48)$$

Therefore:

$$R(H_S, \delta) = \sum_{i=1}^{N} \sum_{j=1}^{N} R(s_{i,j}; \varphi_{i,j}) \tag{6.49}$$

From (6.47), one has

$$\sum_{Q \in \mathscr{G}} w(S, Q) P(\delta^{Sg}(x) = d_Q/H_S) \leq \sum_{Q \in \mathscr{G}} w(S', Q) P(\delta^{Sg}(x) = d_Q/H_S), \quad \forall S, S' \in \mathscr{G}$$
$$\tag{6.50}$$

Thus, the multiple testing statistical procedure δ^{Sg} is unbiased.

Third, we prove that the procedure (6.42) is optimal in the class of unbiased statistical procedures for the threshold graph identification in the sign similarity network. Let $\delta(x)$ be another unbiased statistical procedure for the threshold graph identification in the sign similarity network. Then $\delta(x)$ generates a partition of the sample space $R^{N \times n}$ into $L = 2^M$, $M = N(N-1)/2$ parts:

$$D_G = \{x \in R^{N \times n} : \delta(x) = d_G\}; \quad \bigcup_{G \in \mathscr{G}} D_G = R^{N \times n}$$

Define

$$\begin{aligned} A_{i,j} &= \bigcup_{G: g_{i,j}(x)=0} D_G \\ \overline{A_{i,j}} &= \bigcup_{G: g_{i,j}(x)=1} D_G \end{aligned} \tag{6.51}$$

and

$$\varphi_{i,j}(x) = \begin{cases} 0, & x \in A_{i,j} \\ 1, & x \notin A_{i,j} \end{cases} \tag{6.52}$$

The tests (6.52) are the tests for individual hypotheses testing (6.44). Since the procedure $\delta(x)$ is unbiased, one has

$$\sum_{Q \in \mathscr{G}} w(S, Q) P(\delta(x) = d_Q/H_S) \leq \sum_{Q \in \mathscr{G}} w(S', Q) P(\delta(x) = d_Q/H_S), \quad \forall S, S' \in \mathscr{G}$$

Consider the hypotheses H_S and $H_{S'}$ which differ only in two components $s_{i,j} \neq s'_{i,j}$; $s_{j,i} \neq s'_{j,i}$. Taking into account the unbiasedness of the procedure δ and the structure of the loss function (4.6), one has $R(s_{i,j}, \varphi_{i,j}) \leq R(s'_{i,j}, \varphi_{i,j})$. This means that two decision tests (6.52) are unbiased. Therefore,

$$P(\varphi_{i,j} = 1/p_0) = \alpha_{i,j} = \frac{b_{i,j}}{a_{i,j} + b_{i,j}}.$$

Since we are restricted to the tests based only on $u_i(t)$, $u_j(t)$ and the test $\varphi_{i,j}^{Sg}$ is UMP among tests of this class at the significance level $\alpha_{i,j}$, for any test $\varphi_{i,j}$ based only on $u_i(t)$, $u_j(t)$ one has:

$$R(s_{i,j}, \varphi_{i,j}^{Sg}) \leq R(s_{i,j}, \varphi_{i,j})$$

From (6.49), one has

$$R(H_S, \delta^{Sg}) \leq R(H_S, \delta)$$

for any adjacency matrix S. The theorem is proved.

Note: In the proof of the theorem, we restricted ourselves in nonrandomized tests only. The generalization to the overall case easily could be obtained.

Chapter 7
Applications to Market Network Analysis

Abstract In this chapter, we use the general approach developed in previous chapters to analyze uncertainty of market network structure identification. The results of the Chap. 4 are illustrated by numerical experiments using the data from different stock markets. We study uncertainty of identification algorithms of the following market network structures: market graph, maximum cliques and maximum independent sets in the market graph, maximum spanning tree, and planar maximally filtered graph. Uncertainties of identification of different network structures are compared on the base of risk function for associated multiple decision statistical procedures.

7.1 Market Network Analysis

Network models of financial markets have attracted attention during the last decades [4–7, 39, 66, 87]. A common network representation of the stock market is based on Pearson correlations stock's returns. In such a representation, each stock corresponds to a vertex and a weight of edge between two vertices is estimated by sample Pearson correlation of corresponding returns. In our setting, the obtained complete weighted graph corresponds to a sample Pearson correlation network. The obtained network is a complete weighted graph. In order to simplify the network and preserve the key information, different filtering techniques are used in the literature.

One of the filtering procedures is the extraction of a minimal set of important links associated with the highest degree of similarity belonging to the maximum spanning tree (MST) [66]. The MST was used to find a topological arrangement of stocks traded in a financial market which has associated with a meaningful economic taxonomy. This topology is useful in the theoretical description of financial markets and in search of economic common factors affecting specific groups of stocks. The topology and the hierarchical structure associated to it is obtained by using information present in the time series of stock prices only.

© The Author(s) 2020
V. A. Kalyagin et al., *Statistical Analysis of Graph Structures in Random Variable Networks*, SpringerBriefs in Optimization,
https://doi.org/10.1007/978-3-030-60293-2_7

The reduction to a minimal skeleton of links leads to loss of valuable information. To overcome this issue in [87], it was proposed to extent the MST by iteratively connecting the most similar nodes until the graph can be embedded on a surface of a given genus k. For example, for $k = 0$ the resulting graph is planar, which is called Planar Maximally Filtered Graph (PMFG). It was concluded in [87] that the method is pretty efficient in filtering relevant information about the connection structure both of the whole system and within obtained clusters.

Another filtering procedure leads to the concept of market graph [4–6]. A market graph (MG) is obtained from the original network by removing all edges with weights less than a specified threshold $\gamma_0 \in [-1, 1]$. Maximum cliques (MC) and maximum independent sets (MIS) analysis of the market graph was used to obtain valuable knowledge about the structure of the stock market. For example, it was noted in [89] that the peculiarity of the Russian market is reflected by the strong connection between the volume of stocks and the structure of maximum cliques. The core set of stocks of maximum cliques for Russian market is composed by the most valuable stocks. These stocks account for more than 90% of the market value and represent the largest Russian international companies from banking and natural resource sectors. In contrast, the core set of stocks of the maximum cliques for the US stock market has a different structure without connection to the stocks values.

Today network analysis of financial markets is a very active area of investigation and various directions are developed in order to obtain valuable information for different stock markets [67]. Most of publications are related with numerical algorithms and economic interpretations of obtained results. Much less attention is paid to uncertainty of obtained results generated by stochastic nature of the market. In this chapter, we apply the results of the Chap. 4

Let N be a number of stocks, n be a number of days of observations. In our study, financial instruments are characterized by daily returns of the stocks. Stock k return for day t is defined as

$$R_k(t) = \ln \frac{P_k(t)}{P_k(t-1)}, \tag{7.1}$$

where $P_k(t)$ is the adjusted closing price of stock k on day t. We assume that for fixed k, $R_k(t)$, $t = 1, \ldots, n$, are independent random variables with the same distribution as R_k (i.i.d.) and the random vector $R = (R_1, \ldots, R_N)$ has multivariate distribution with correlation matrix

$$\Gamma = (\rho_{ij}) = \begin{pmatrix} \rho_{1,1} & \cdots & \rho_{1,N} \\ \cdots & \cdots & \cdots \\ \rho_{N,1} & \cdots & \rho_{N,N} \end{pmatrix}. \tag{7.2}$$

For this model, we introduce the reference (true) network model, which is a complete weighted graph with N nodes and weight matrix $\Gamma = (\rho_{i,j})$. For the reference network, one can consider corresponding reference structures, for example, reference MST, reference PMFG, reference market graph, and others.

Let $r_k(t), k = 1, \ldots, N, t = 1, \ldots, n$, be the observed values of returns. Define the sample covariance

$$s_{i,j} = \frac{1}{n-1} \sum_{t=1}^{n} (r_i(t) - \bar{r}_i)(r_j(t) - \bar{r}_j),$$

and sample correlation

$$r_{i,j} = \frac{s_{i,j}}{\sqrt{s_{i,i} s_{j,j}}}$$

where $\bar{r}_i = (1/n) \sum_{t=1}^{n} r_i(t)$. Using the sample correlations we introduce the sample network, which is a complete weighted graph with N nodes and weight matrix $\hat{\Gamma} = (r_{i,j})$. For the sample network, one can consider the corresponding sample structures, for example, sample MST, sample PMFG, sample market graph, and others.

7.2 Measures of Uncertainty

To handle statistical uncertainty, we propose to compare the sample network with the reference network. Our comparison will be based on the risk function connected with the loss function developed in the Chap. 4.

For a given structure $\mathscr{S}$, we introduce a set of hypothesis:

- $h_{i,j}$: edge between vertices i and j is not included in the reference structure $\mathscr{S}$;
- $k_{i,j}$: edge between vertices i and j is included in the reference structure $\mathscr{S}$.

To measure the losses, we consider two types of errors:

Type I error or False Positive error: edge is included in the sample structure when it is absent in the reference structure;
Type II error or False Negative error: edge is not included in the sample structure when it is present in the reference structure.

Let $a_{i,j}$ be the loss associated with the error of the first kind and $b_{i,j}$ the loss associated with the error of the second kind for the edge (i, j). According to the Chap. 4, we define the risk for additive loss function for a given structure $\mathscr{S}$, given identification procedure δ, and given number of observations n as

$$\mathscr{R}(\mathscr{S}; \delta, n) = \sum_{1 \le i < j \le N} [a_{i,j} P_n(d_{k_{i,j}} | h_{i,j}) + b_{i,j} P_n(d_{h_{i,j}} | k_{i,j})], \tag{7.3}$$

where $P_n(d_{k_{i,j}} | h_{i,j})$ is the probability of rejecting hypothesis $h_{i,j}$ when it is true and $P_n(d_{h_{i,j}} | k_{i,j})$ is the probability of accepting hypothesis $h_{i,j}$ when it is false. In our case, errors can be defined using a number of edges different in sample structure with respect to reference structure and vice versa.

Let

$$
x_1^{i,j} = \begin{cases} 1, \text{ if edge } (i, j) \text{ is included in the sample structure,} \\ \quad \text{but is not present in the reference structure} \\ 0, \text{ otherwise,} \end{cases}
$$

and

$$
x_2^{i,j} = \begin{cases} 1, \text{ if edge } (i, j) \text{ is not included in the sample structure,} \\ \quad \text{but is present in the reference structure} \\ 0, \text{ otherwise.} \end{cases}
$$

Define

$$
X_1 = \sum_{1 \le i < j \le N} x_1^{i,j}; \quad X_2 = \sum_{1 \le i < j \le N} x_2^{i,j}.
$$

X_1 is the number of Type I errors (FP), X_2 is the number of Type II errors (FN).

Let us now define the random variable

$$
X = \frac{1}{2} \left(\frac{X_1}{M_1} + \frac{X_2}{M_2} \right), \tag{7.4}
$$

where M_1 is a maximal possible value of X_1, and M_2 is a maximal possible value of X_2. Random variable $X \in [0, 1]$ describes total fraction of errors. Indeed $X = (1/2)(\text{FPR} + \text{FNR})$, where FPR is the False Positive Rate, and FNR is the False Negative Rate. Let $\mathscr{E}(\mathscr{S}; \delta, n) = E(X)$. We define the *measure of statistical uncertainty* of level $\mathscr{E}_0$ of the procedure δ for the structure $\mathscr{S}$ as the minimal number of observations $n(\mathscr{E}_0)$ such that

$$
\mathscr{E}(\mathscr{S}; \delta, n) \le \mathscr{E}_0,
$$

where $\mathscr{E}_0$ is a given positive number.

Note that the $\mathscr{E}$-measure of statistical uncertainty has natural interpretation and coincides with risk function. Namely, if M_1 and M_2 are constants (nonrandom) then

$$
\mathscr{E}(\mathscr{S}; \delta, n) = \sum_{1 \le i < j \le N} \left[\frac{1}{2M_1} P_n(x_1^{i,j} = 1) + \frac{1}{2M_2} P_n(x_2^{i,j} = 1) \right].
$$

Now if in (7.3) we put $a_{i,j} = 1/2M_1$ and $b_{i,j} = 1/2M_2$ then

$$\text{Risk}(\mathscr{S}; \delta, n) = \sum_{1 \leq i < j \leq N} \left[\frac{1}{2M_1} P_n(d_{k_{i,j}} | h_{i,j}) + \frac{1}{2M_2} P_n(d_{h_{i,j}} | k_{i,j}) \right].$$

Since $P_n(d_{k_{i,j}} | h_{i,j}) = P_n(x_1^{i,j} = 1)$ and $P_n(d_{h_{i,j}} | k_{i,j}) = P_n(x_2^{i,j} = 1)$, it implies that

$$\mathscr{E}(\mathscr{S}; \delta, n) = \text{Risk}(\mathscr{S}; \delta, n), \quad \text{and} \quad n_{\mathscr{E}} = n_{\mathscr{R}}.$$

In particular, the described above relation between risk function and $\mathscr{E}$-measure takes place for such structures as MG, MST, PMFG.

7.3 Numerical Experiments Framework

In this section, applications of described measures of statistical uncertainty are illustrated by numerical experiments in the framework of Gaussian network model of stock market. By Gaussian network model of stock market, we understand the following model: for a fixed stock k, its returns $R_k(t), t = 1, \ldots, n$, are independent random variables with the same distribution as R_k (i.i.d.) and the random vector $R = (R_1, \ldots, R_N)$ has multivariate Gaussian distribution.

In our experiments, we consider the network with $N = 250$ nodes. Vector of returns R has Gaussian distribution, $R \sim N(0, \Gamma)$, where the correlation matrix $\Gamma = (\rho_{i,j})$ consists of pairwise correlations of daily returns of a set of 250 financial instruments traded in the US NYSE stock markets over a period of 365 consecutive trading days in 2010–2011. We use the matrix $\Gamma = (\rho_{i,j})$ as a weight matrix for our reference network. To construct the sample network, we generate the sample $r(t), t = 1, 2, \ldots, n$ from multivariate normal distribution $N(0, \Gamma)$. To measure statistical uncertainty $n(\mathscr{E}_0)$ of statistical procedure δ for the network structure $\mathscr{S}$, we use $\mathscr{E}(\mathscr{S}; \delta, n)$ which we estimate in the following way:

1. In the reference network find reference structure $\mathscr{S}$.
2. Generate sample $r(t), t = 1, 2, \ldots, n$.
3. Calculate sample correlations $r_{i,j}$.
4. In sample network (with weight matrix $\hat{\Gamma}$) find the sample structure $\hat{\mathscr{S}}$.
5. Calculate fraction of errors of type I (X_1/M_1), fraction of errors of type II (X_2/M_2) and total fraction of errors (X) by (7.4).
6. Repeat steps 1–5 and calculate mean of X which is an estimation of $\mathscr{E}(\mathscr{S}, n)$.

Table 7.1 Statistical
uncertainty of MST

$\mathcal{E}_0$	0,5	0,4	0,3	0,2	0,1
$n(\mathcal{E}_0)$	300	500	1000	3000	10,000

7.4   Statistical Uncertainty of MST

Observe that if $X = 0$ then reference MST, and sample MST are equal; $X = 1$
means that reference MST and sample MST are completely different, i.e., have no
common edges. The latter situation may hold for several sample MSTs under fixed
reference MST. For maximum spanning tree, one has $M_1 = M_2 = N - 1$, where
N is a number of vertices in considered network. The number of edges in any MST
is fixed and equal $(N - 1)$; therefore, the number of Type I errors is equal to the
number of Type II errors, $X_1 = X_2$. Since M_1 and M_2 are constants, the statistical
uncertainty of identification procedure δ for MST identification can be defined from
the inequality:

$$\frac{1}{2(N-1)} \sum_{1 \leq i < j \leq N} [P_n(x_1^{i,j}=1) + P_n(x_2^{i,j}=1)] = \frac{1}{(N-1)} \sum_{1 \leq i < j \leq N} P_n(x_1^{i,j}=1) \leq \mathcal{E}_0.$$

The results of the study of statistical uncertainty of Kruskal algorithm for MST
identification are presented in the Table 7.1.

In all tables we make an rounding of $n(\mathcal{E}_0)$ to the closest number of the form
$k 10^s, k = 1, 2, \dots, 9$. As one can see for Kruskal algorithms for MST identification,
the condition $\mathcal{E}(\text{MST}; \delta, n) \leq 0.1$ is achieved when the number of observations n is
larger than 10 000. Note that when $n = 1000$ sample and reference MSTs have only
70% of common edges. Moreover by further increasing the number of observations
does not lead to considerable decrease of statistical uncertainty of MST.

7.5   Statistical Uncertainty of PMFG

Observe that $X = 0$ means that reference PMFG and sample PMFG are equal;
$X = 1$ means that reference PMFG and sample PMFG are completely different, i.e.,
have no common edges. The latter situation may hold for several sample PMFGs
under fixed reference PMFG. For planar maximally filtered graph, one has $M_1 =
M_2 = 3N-6$, where N is a number of vertices in considered network (see [87]). The
number of edges in any PMFG is fixed and equal $(3N - 6)$; therefore, the number
of Type I errors is equal to the number of Type II errors, $X_1 = X_2$. Since M_1 and
M_2 are constants, the uncertainty of statistical procedure δ for PMFG identification
can be defined from the inequality:

Table 7.2 Statistical
uncertainty of PMFG

$\mathscr{E}_0$	0,5	0,4	0,3	0,2	0,1
$n(\mathscr{E}_0)$	400	1000	3000	10,000	40,000

$$\frac{1}{(3N-6)} \sum_{1 \le i < j \le N} P_n(x_1^{ij} = 1) \le \mathscr{E}_0.$$

For PMFG identification we use Kruskal type algorithm suggested in [87]: a list of edges is sorted in descending order according to the weight and following the ordered list an edge is added to the PMFG if and only if obtained after this operation graph is planar. This algorithm has polynomial computational complexity. The results of the study of statistical uncertainty of Kruskal type algorithm for PMFG identification are presented in Table 7.2. In all tables, we make an rounding of $n(\mathscr{E}_0)$ to the closest number of the form $k10^s$, $k = 1, 2, \ldots, 9$. As one can see for Kruskal algorithms for PMFG identification the condition $\mathscr{E}(\text{PMFG}, n) \le 0.2$ is achieved only when the number of observations n is larger than $10\,000$. This is worse than for MST identification. Moreover, the same as for the MST, by further increasing the number of observations, does not lead to considerable decrease of statistical uncertainty of PMFG.

7.6   Statistical Uncertainty of MG

Observe that $X = 0$ means that reference MG and sample MG are equal; $X = 1$ means that sample MG is complement to reference MG. Let us pay attention that the latter situation for market graph is possible in only one case, in contrast to MST. For the market graph, one has $M_1 = (1/2)N(N-1) - M$, $M_2 = M$, where N is the number of the vertices in the considered network and M is the number of edges in the given reference market graph. Since M_1 and M_2 are constants, the uncertainty of statistical procedure δ for MG identification can be defined from the inequality:

$$\frac{1}{2} \sum_{1 \le i < j \le N} \left[\frac{1}{M_1} P_n(x_1^{i,j} = 1) + \frac{1}{M_2} P_n(x_2^{i,j} = 1) \right] \le \mathscr{E}_0.$$

The results of the study of statistical uncertainty of simultaneous inference procedure δ^S with $\alpha_{i,j} = 0, 5$ for MG identification are presented in the Table 7.3. In all tables, we round $n(\mathscr{E}_0)$ to the closest number of the form $k10^s$, $k = 1, 2, \ldots, 9$. As one can see the condition $\mathscr{E}(\text{MG}; \delta^S, n) \le 0.1$ is achieved under the number of observations $n = 300$ for all thresholds γ_0, which is much more reasonable than the statistical uncertainty of MST and PMFG. Note that uncertainty of MG identification has a small variation with the change of threshold.

Table 7.3 Statistical
uncertainty of MG

$\mathcal{E}_0$		0,3	0,15	0,1	0,05
$n(\mathcal{E}_0)$, $\gamma_0 = 0, 1$		10	100	300	1000
$n(\mathcal{E}_0)$, $\gamma_0 = 0, 3$		10	100	300	1000
$n(\mathcal{E}_0)$, $\gamma_0 = 0, 5$		10	100	300	1000
$n(\mathcal{E}_0)$, $\gamma_0 = 0, 7$		10	100	300	1000

7.7 Statistical Uncertainty of MC

Since a graph may contain many maximum cliques, in our experiments we choose the maximum clique with maximal weight (MCMW). The weight of a clique in a market graph is the sum of weights of corresponding edges in the network. In each of all our Experiments, there was only one maximum clique of maximal weight. We slightly modify the definitions of the numbers of errors X_1, X_2. X_1 is the number of edges in the sample MCMW which are absent in the reference market graph. X_2 is the number of edges in the reference MCMW, which are absent in the sample market graph. For maximum clique, one has $M_1 = C_s(C_s - 1)/2$, $M_2 = C_r(C_r - 1)/2$, where C_s is a number of vertices in the sample MCMW and C_r is a number of vertices in the reference MCMW. Since C_s and therefore, M_1 are random variables, the measure of statistical uncertainty for MCMW can be defined from the inequality:

$$\frac{1}{2}E_n\left(\frac{X_1}{M_1}\right) + \frac{1}{2}\sum_{1 \le i < j \le N}\frac{P_n(x_2^{ij} = 1)}{M_2} \le \mathcal{E}_0.$$

Observe that $X = 0$ means that reference MCMW and sample MCMW are equal. $X = 1$ means that $X_1 = M_1$ and $X_2 = M_2$. $X_1 = M_1$ corresponds to the situation when all edges of sample MCMW were included incorrectly, i.e., vertices of sample MCMW induce the empty subgraph in the reference market graph. $X_2 = M_2$ corresponds to the situation when all edges of the reference MCMW are absent in a sample market graph, i.e., vertices of the reference MCMW induce empty subgraph in the sample market graph.

For numerical experiments, we consider the following MC identification algorithm: first we apply simultaneous inference procedure for market graph identification and then we apply any exact algorithm for maximum clique identification. Results of the study of statistical uncertainty of this procedure are presented in the Table 7.4.

As one can see the condition $\mathcal{E}(\text{MCMW}, n) \le 0.1$ is achieved under the number of observations $n_\mathcal{E} = 100$ for all considered thresholds γ_0.

Table 7.4 Statistical
uncertainty of maximum
clique (MCMW)

$\mathscr{E}_0$	0,3	0,2	0,1	0,05
$n(\mathscr{E}_0), \gamma_0 = 0, 5$	2	10	20	40
$n(\mathscr{E}_0), \gamma_0 = 0, 6$	5	20	40	100
$n(\mathscr{E}_0), \gamma_0 = 0, 7$	30	50	100	300
$n(\mathscr{E}_0), \gamma_0 = 0, 8$	20	30	60	130

Table 7.5 Statistical
uncertainty of maximum
independent set (MISMW)

$\mathscr{E}_0$	0,3	0,2	0,1
$n(\mathscr{E}_0), \gamma_0 = 0, 000$	80	200	500
$n(\mathscr{E}_0), \gamma_0 = 0, 050$	50	170	600
$n(\mathscr{E}_0), \gamma_0 = 0, 075$	40	130	500
$n(\mathscr{E}_0), \gamma_0 = 0, 100$	20	80	370

7.8   Statistical Uncertainty of MIS

Since a graph may contain many maximum independent sets, in our experiments we
choose the maximum independent set with minimal weight (MISMW). The weight
of an independent set in a market graph is the sum of weights of corresponding
edges in the network. In each of our experiments, there was only one maximum
independent set of minimal weight. For maximum independent set, one has $M_1 =
I_r(I_r - 1)/2$, $M_2 = I_s(I_s - 1)/2$, where I_r is a number of vertices in the reference
MIS and I_s is a number of vertices in a sample MISMW. Since I_s and therefore, M_2
are random variables, the measure of statistical uncertainty for MISMW is defined
from the inequality:

$$\frac{1}{2} \sum_{1 \leq i < j \leq N} \frac{P_n(x_1^{ij} = 1)}{M_1} + \frac{1}{2} E_n \left(\frac{X_2}{M_2} \right) \leq \mathscr{E}_0.$$

Observe that $X = 0$ means that reference MISMW and sample MISMW are
equal. $X = 1$ means that $X_1 = M_1$ and $X_2 = M_2$. $X_1 = M_1$ corresponds to
the situation when vertices of the reference MISMW induce complete subgraph in
a sample market graph. $X_2 = M_2$ corresponds to the situation when vertices of a
sample MISMW induce complete subgraph in the reference market graph.

For numerical experiments, we consider the following MIS identification algorithm: first we apply simultaneous inference procedure for market graph identification and then we apply any exact algorithm for maximum independent set
identification. Results of the study of statistical uncertainty of this procedure are
presented in the Table 7.5.

As one can see the condition $\mathscr{E}(\text{MISMW}, n) \leq 0.1$ is achieved under the number
of observations $n_\mathscr{E} = 600$ for all considered thresholds.

Chapter 8
Conclusion

In this book, we presented the first steps to investigate uncertainty of network structure identification in random variable networks. We propose to measure uncertainty as a risk (expected value of losses) associated with a loss function. On this basis we introduce the concepts of robustness and optimality of identification procedures. The main results of our investigation described in the book are the following:

- It is discovered that for the elliptical probabilistic network model, different correlation networks and associated network structures are closely related.
- New class of identification algorithms based on a new measure of similarity (sign similarity) is introduced and investigated. It is proved that these identification algorithms are robust (distribution free) in the class of elliptical distributions.
- Optimality of simultaneous inference identification procedure is investigated for the case of additive loss function. We prove optimality of simultaneous inference for concentration graph identification in Gaussian partial correlation network in the class of unbiased procedures. For threshold graph identification, we prove optimality of simultaneous inference in Gaussian Pearson correlation network in the class of statistical procedures invariant under scale/shift transformations. Optimality of simultaneous inference for threshold graph identification in sign similarity network is proved under some additional constraints.
- It is shown that popular market network structures essentially differ by the level of uncertainty of its identification.

Our results are not complete and open a large area for further investigations.

© The Author(s) 2020
V. A. Kalyagin et al., *Statistical Analysis of Graph Structures in Random Variable Networks*, SpringerBriefs in Optimization,
https://doi.org/10.1007/978-3-030-60293-2_8

References

1. Anderson, T.W.: An Introduction to Multivariate Statistical Analysis, 3rd edn. Wiley-Interscience, New York (2003)
2. Benjamini, Y., Hochberg, Y.: Controlling the false discovery rate: a practical and powerful approach to multiple testing. J. R. Stat. Soc. Ser. B (Methodological) **57**(1):289–300 (1995)
3. Blomqvist, N.: On a measure of dependence between two random variables. Ann. Math. Stat. **21**:593–600 (1950)
4. Boginski, V., Butenko, S., Pardalos, P.M.: Mining market data: a network approach. Comput. Oper. Res. **33**(11):3171–3184 (2006)
5. Boginski, V., Butenko, S., Pardalos, P.M.: On structural properties of the market graph. Innov. Financ. Econ. Netw. 29–45 (2003)
6. Boginski, V., Butenko, S., Pardalos, P.M.: Statistical analysis of financial networks. Comput. Stat. Data Anal. **48**(2):431–443 (2005)
7. Bonanno, G.: Networks of equities in financial markets. Eur. Phys. J. B-Condens. Matter Complex Syst. **38**(2):363–371 (2004)
8. Bonanno, G.: Topology of correlation-based minimal spanning trees in real and model markets. Phys. Rev. E **68**:046130 (2003)
9. Bretz, F., Hothorn, T., Westfall, P.: Multiple Comparisons Using R. Taylor and Francis Group, Boca Raton (2011)
10. Brugere, I., Gallagher, B., Tanya, Y.B.-W.: Network structure inference, a survey: motivations, methods, and applications. ACM Comput. Surv. (CSUR) **51**(2):1(39) (2018)
11. Bullmore, E., Sporns, O.: Complex brain networks: graph theoretical analysis of structural and functional systems. Nat. Rev. Neurosci. **10**(3):186(198) (2009)
12. Butte, A.J., Kohane, I.S.: Mutual information relevance networks: functional genomic clustering using pairwise entropy measurements. In: Biocomputing 2000, Proceedings of the Pacific Symposium, pp. 418–429 (2000)
13. Cayley, A.: A theorem on trees. Q. J. Pure Appl. Math. **23**:376–378 (1889)
14. Chu, J., Nadarajah, S.: A statistical analysis of UK financial network. Phys. A Stat. Mech. Appl. **471**:445–459 (2017)
15. Carraghan, R., Pardalos, P.: An exact algorithm for the maximum clique problem. Oper. Res. Lett. **9**(6):375–382 (1990)
16. Dempster, A.P.: Covariance selection. Biometrics **28**:157–175 (1972)
17. Drton, M., Maathuis, M.: Structure learning in graphical modeling. Ann. Rev. Stat. Appl. **4**:365–393 (2017)

© The Author(s) 2020
V. A. Kalyagin et al., *Statistical Analysis of Graph Structures in Random Variable Networks*, SpringerBriefs in Optimization,
https://doi.org/10.1007/978-3-030-60293-2

18. Drton, M., Perlman, M.: A SINful approach to Gaussian graphical model selection. J. Stat. Plann. Inference **138**:1179–1200 (2008)
19. Drton, M., Perlman, M.: Model selection for Gaussian concentration graph. Biometrika **91**(3):591–602 (2004)
20. Drton, M., Perlman, M.: Multiple testing and error control in Gaussian graphical model selection. Stat. Sci. **22**(3):430–449 (2008)
21. Edwards, D.: Introduction to Graphical Modeling. Springer, New York (2000)
22. Emmert-Streib, F., Dehmer, M.: Identifying critical financial networks of the DJIA: towards a network based index. Complexity **16**(1):24–33 (2010)
23. Emmert-Streib, F., Dehmer, M.: Influence of the time scale on the construction of financial networks. PLoS One **5**(9):e12884 (2010)
24. Eoma, C.: Topological properties of stock networks based on minimal spanning tree and random matrix theory in financial time series. Phys. A Stat. Mech. Appl. **388**:900–906 (2009)
25. Fang, H.-B., Fang, K.-T., Kotz, S.: The meta-elliptical distrivutions with given marginals. J. Multivar. Anal. **82**:445–459 (2017)
26. Galazka, M.: Characteristics of the polish stock market correlations. Int. Rev. Fin. Anal. **20**(1):1–5 (2011)
27. Garas, A., Argyrakis, P.: Correlation study of the Athens stock exchange. Phys. A Stat. Mech. Appl. **380**:399–410 (2007)
28. Goldengorin, B., Kocheturov, A., Pardalos, P.M.: A pseudo-boolean approach to the market graph analysis by means of the p-median model. In: Aleskerov, F., et al. (eds.) Clusters, Orders, and Trees: Methods and Applications. In: Honor of Boris Mirkin's 70th Birthday, Springer Optimization and Its Applications, Vol. 92, pp. 77–89 (2014)
29. Gross, J., Yellen, J.: Graph Theory and Its Applications. CRC Press, Boca Raton (2006)
30. Gunawardena, A.D.A.: Optimal selection of an independent set of cliques in a market graph. Int. Proc. Econ. Dev. Res. **29**:281–285 (2012)
31. Guo, X., Zhang, H., Tian, T.: Development of stock correlation networks using mutual information and financial big data. PLoS One **13**(4):e0195941 (2018)
32. Gupta, F.K., Varga, T., Bodnar, T.: Elliptically Contoured Models in Statistics and Portfolio Theory. Springer, New York (2013)
33. Hero, A., Rajaratnam, B.: Hub discovery in partial correlation graphs. IEEE Trans. Inf. Theory **58**(9):6064–6078 (2012)
34. Hochberg, Y., Tamhane, A.: Multiple Comparison Procedures. Wiley, New York (1987)
35. Hochberg, Y.A.: Sharper Bonferroni procedure for multiple tests of significance. Biometrika **75**(4):800–802 (1988)
36. Hoeffding, W.: On the distribution of the rank correlation coefficient τ when the variates are not independent. Biometrika **34**:183–196 (1947)
37. Holm, S.: A simple sequentially rejective multiple test procedure. Scand. J. Stat. **6**(2):65–70 (1979)
38. Horvath, S.: Weighted Network Analysis. Applications in Genomics and Systems Biology. Springer Book. ISBN 978-1-4419-8818-8 (2011)
39. Huang, W.-Q., Zhuang, X.-T., Yao, S.: A network analysis of the Chinese stock market. Phys. A Stat. Mech. Appl. **388**(14):2956–2964 (2009)
40. Jallo, D.: Network-based representation of stock market dynamics: an application to American and Swedish stock markets. In: Goldengorin, B., Kalyagin, V., Pardalos, P.M., (eds.) Models, Algorithms, and Technologies for Network Analysis. In: Springer Proceedings in Mathematics and Statistics. **32**:93–106 (2013)
41. Jordan, M.I.: Graphical models. Stat. Sci. **19**:140–155 (2004)
42. Jung, W.-S.: Characteristics of the Korean stock market correlations. Phys. A Stat. Mech. Appl. **361**:263–271 (2006)
43. Kalyagin, V.A., Koldanov, A.P., Koldanov, P.A., Zamaraev, V.A.: Market graph and Markowitz model. Optimization in Science and Engineering (In Honor of the 60th Birthday of Panos M. Pardalos). Springer Science, Business Media, pp. 301–313 (2014)

44. Kalyagin, V.A., Koldanov, A.P., Koldanov, P.A., Pardalos, P.M., Zamaraev, V.A.: Measures of uncertainty in market network analysis. Phys. A Stat. Mech. Appl. **413**(1):59–70 (2014)
45. Kapteyn, J.C.: Definition of the correlation coefficient. Mon. Not. R. Astron. Soc. **72**:518–525 (1912)
46. Kazakov, M., Kalyagin, V.: Spectral properties of financial correlation matrices. In: Kalyagin, V., Koldanov, P., Pardalos, P. (eds.) Models, Algorithms and Technologies for Network Analysis, pp. 135–156. Springer, Cham (2016)
47. Kenett, D.Y.: Dominating clasp of the financial sector revealed by partial correlation analysis of the stock market. PLoS One **5**(12):e15032 (2010)
48. Keskin, V., Deviren, B., Kocakaplan, Y.: Topology of the correlation networks among major currencies using hierarchical structure methods. Phys. A Stat. Mech. Appl. **390**:719–730 (2011)
49. Khare, K., Sang-Yun, O., Rajaratnam, B.: A convex pseudo-likelihood framework for high dimensional partial correlation estimation with convergence guarantees. J. R. Stat. Soc. Ser. B Stat. Methodol. **77**:803825 (2015)
50. Kim, D.Y., Jeong, H.: Systematic analysis of group identification in stock markets. Phys. Rev. **72**:046133 (2005)
51. Kim, H.J.: Scale-free networks in financial correlations. J. Phys. Soc. Jpn. **71**:1–5 (2002)
52. Kocheturov, A., Batsyn, M., Pardalos, P.M.: Dynamics of cluster structures in a financial market network. Phys. A Stat. Mech. Appl. **413**:523–533 (2014)
53. Kocheturov, A., Batsyn, M., Pardalos, P.M.: Dynamics of cluster structures in stock market networks. J. New Econ. Assoc. **28**(4):12–30 (2015)
54. Koldanov, A.P., Koldanov, P.A., Kalyagin, V.A., Pardalos, P.M.: Statistical procedures for the market graph construction. Comput. Stat. Data Anal. **68**:17–29 (2013)
55. Koldanov, P.A., Koldanov, A.P., Kalyagin, V.A., Pardalos, P.M.: Uniformly most powerful unbiased test for conditional independence in Gaussian graphical model. Stat. Probab. Lett. **122**:90–95 (2017)
56. Korenkevych, D., Skidmore, F., Goldengorin, B., Pardalos, P.M.: How close to optimal are small world properties of human brain networks? In: Goldengorin, et al. (eds.) Models, Algorithms, and Technologies for Network Analysis. Springer Proceedings in Mathematics and Statistics, Vol. 32, pp. 117–127 (2013)
57. Kruskal, W.H.: Ordinal measures of association. J. Am. Stat. Assoc. **53**:814–861 (1958)
58. Lauritzen, S.L., Sheehan, N.A.: Graphical models for genetic analysis. Stat. Sci. **18**(3):489–514 (2003)
59. Lauritzen, S.L.: Graphical Models. Oxford University Press, Oxford (1996)
60. Lehmann, E.L.: A theory of some multiple decision problems, I. Ann. Math. Stat. **28**:1–25 (1957)
61. Lehmann, E.L., Romano, J.P.: Generalizations of the familywise error rate. Ann. Stat. **33**:1138–1154 (2005)
62. Lehmann, E.L., Romano, J.P.: Testing Statistical Hypotheses. Springer, New York (2005)
63. Li, S., He, J., Zhuang, Y.: A network model of the interbank market. Phys. A Stat. Mech. Appl. **389**:5587–5593 (2010)
64. Liang, A., Song, O., Qiu, P.: An equivalent measure of partial correlation coefficients for high-dimensional Gaussian graphical models. J. Am. Stat. Assoc. **110**(511), 1248–1265 (2015)
65. Lindskog, F., McNeil, A., Schmock, U.: Kendall tau for elliptical distributions. In: Bol, G., et al. (eds.) Credit Risk: Measurement, Evaluation and Management, pp. 149–156. Physica-Verlag, New York (2003)
66. Mantegna, R.N.: Hierarchical structure in financial markets. Eur. Phys. J. B-Condens. Matter Complex Syst. **11**(1):193–197 (1999)
67. Martia, G., Nielsen, F., Bikowski, M., Donnat, F.: A review of two decades of correlations, hierarchies, networks and clustering in financial markets. arXiv:1703.00485v1 (2017)
68. Meinshausen, N., Buhlmann, P.: High-dimensional graphs and variable selection with the lasso. Ann. Stat. **34**:14361462 (2006)

69. Micciche, S., Bonanno, G., Lillo, F., Mantegna, R.N.: Degree stability of a minimum spanning tree of price return and volatility. Phys. A Stat. Mech. Appl. **324**(1–2):66–73 (2003)

70. Namaki, A., Jafari, G.R., Raei, R.: Comparing the structure of an emerging market with a mature one under global perturbation. Phys. A Stat. Mech. Appl. **390**(17):3020–3025 (2011)

71. Namaki, A., Shirazi, A.H., Jafari, G.R., Raei, R.: Network analysis of a financial market based on genuine correlation and threshold method. Phys. A Stat. Mech. Appl. **390**(17):3835–3841 (2011)

72. Nguyen, Q.: One-factor model for cross-correlation matrix in the Vietnamese stock market. Phys. A Stat. Mech. Appl. **392**(13):2915–2923 (2013)

73. Onnela, J.-P.: Dynamics of market correlations: taxonomy and portfolio analysis. Phys. Rev. E **68**(5):56–110 (2003)

74. Onnela, J.-P., Kaski, K., Kertesz, J.: Clustering and information in correlation based financial networks. Eur. Phys. J. B Condens. Matter Complex Syst. **38**(2):353–362 (2004)

75. Peng, J., Wang, P., Zhou, N., Zhu, J.: Partial correlation estimation by joint sparse regression models. J. Am. Stat. Assoc. **104**(486):735746 (2009)

76. Plerou, V.: Universal and nonuniversal properties of cross correlations in financial time series. Phys. Rev. **83**:1471–1474 (1999)

77. Rajaratnam, B., Massam, H., Carvalhob, C.: Flexible covariance estimation in graphical models. Ann. Stat. **36**(6):2818–2849 (2008)

78. Ren, Z.: Asymtotic normality and optimalities in estimation of large Gaussian graphical models. Ann. Stat. **43**(3):991–1026 (2015)

79. Sarkar, S.K., Chung-Kuei, C.: The Simes method for multiple hypothesis testing with positively dependent test statistics. J. Am. Stat. Assoc. **92**(440):1601–1608 (1997)

80. Sensoya, A., Tabak, B.M.: Dynamic spanning trees in stock market networks: the case of Asia-Pacific. Phys. A Stat. Mech. Appl. **414**:387–402 (2014)

81. Shapira, Y., Kenett, D.Y., Ben-Jacob, E.: The index cohesive effect of stock market correlations. J. Phys. B **72**(4):657–669 (2009)

82. Shirokikh, O., Pastukhov, G., Boginski, V., Butenko, S.: Computational study of the US stock market evolution: a rank correlation-based network model. Comput. Manag. Sci. **10**(2–3):81–103 (2013)

83. Song, W.-M., Di Matteo, T., Tomaso, A.: Hierarchical information clustering by means of topologically embedded graphs. PLoS One 1–16 (2002)

84. Tabak, B.M., Thiago, R.S., Cajueiro, D.O.: Topological properties of stock market networks: the case of Brazil. Phys. A Stat. Mech. Appl. **389**:3240–3249 (2010)

85. Tse, C.K., Liu, J., Lau, F.C.M.: A network perspective of the stock market. J. Emp. Fin. **17**:659–667 (2010)

86. Tsonis, A.A., Roebber, P.G.: The architecture of the climate network. Phys. A Stat. Mech. Appl. **333**:497(504) (2004)

87. Tumminello, M., Aste, T., Di Matteo, T., Mantegna, R.N.: A tool for filtering information in complex systems. Proc. Natl. Acad. Sci. U. S. A. **102**(30):10421–10426 (2005)

88. Tumminello, M., Lillo, F., Mantegna, R.N.: Correlation, hierarchies and networks in financial markets. J. Econ. Behav. Organ. **752**:40–58 (2010)

89. Vizgunov, A.N., Goldengorin, B., Kalyagin, V.A., Koldanov, A.P., Koldanov, P., Pardalos, P.M.: Network approach for the Russian stock market. Comput. Manag. Sci. **11**:45–55 (2014)

90. Wainwright, M.J., Jordan, M.I.: Graphical models, exponential families and variational inference. Found. Trends Mach. Learn. **1**(1–2):1–305 (2008)

91. Wald, A.: Statistical Decision Functions. Wiley, New York (1950)

92. Wang, G.-J., Xie, C., Chen, Y.-J., Chen, S.: Statistical properties of the foreign exchange network at different time scales: evidence from detrended cross-correlation coefficient and minimum spanning tree. Entropy **15**(5):1643–1662 (2013)

93. Wang, G.-J., Chi, X., Han, F., Sun, B.: Similarity measure and topology evolution of foreign exchange markets using dynamic time warping method: evidence from minimal spanning tree. Phys. A Stat. Mech. Appl. **391**(16):4136–4146 (2012)

94. Wang, G.-J., Xie, C., Shou, C., Yang, J.-J., Yang, M.-Y.: Random matrix theory analysis of cross-correlations in the us stock market: evidence from Pearson correlation coefficient and detrended cross-correlation coefficient. Phys. A Stat. Mech. Appl. **392**:3715–3730 (2013)

95. Zebende, G.F.: DCCA cross-correlation coefficient: quantifying level of cross-correlation. Phys. A Stat. Mech. Appl. **390**:614–618 (2011)

96. Zhang, B., Horvath, S.: A general framework for weighted gene co-expression network analysis, applications in genetics and molecular biology. Stat. Appl. Genet. Mol. Biol. **4**(1):1128 (2005)

97. Zou, Y., Donner, R.V., Marwan, N., Donges, D.F., Kurths, J.: Complex network approaches to nonlinear time series analysis. Phys. Rep. **787**:1(97) (2019)